Metaphorical Circuit

Metaphorical Circuit
Negotiations between
Literature and Science
in Twentieth-Century Japan

Joseph A. Murphy

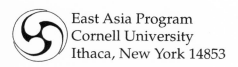
East Asia Program
Cornell University
Ithaca, New York 14853

The Cornell East Asia Series is published by the Cornell University East Asia Program (distinct from Cornell University Press). We publish affordably priced books on a variety of scholarly topics relating to East Asia as a service to the academic community and the general public. Standing orders, which provide for automatic notification and invoicing of each title in the series upon publication, are accepted.

If after review by internal and external readers a manuscript is accepted for publication, it is published on the basis of camera-ready copy provided by the volume author. Each author is thus responsible for any necessary copy-editing and for manuscript formatting. Address submission inquiries to CEAS Editorial Board, East Asia Program, Cornell University, Ithaca, New York 14853-7601.

Number 119 in the Cornell East Asia Series
Copyright © 2004 by Joseph A. Murphy. All rights reserved
ISSN 1050-2955
ISBN 1-885445-19-9 pb
Library of Congress Control Number: 2004102511
Printed in the United States of America
20 19 18 17 16 15 14 13 12 11 10 09 08 07 06 04   9 8 7 6 5 4 3 2 1

Cover image from *Scientific Papers,* by Torahiko Terada, 1936 edition; cover design by Karen K. Smith.

⊖   The paper in this book meets the requirements for permanence of ISO 9706:1994.

For my wife
The thought is insignificant in this case

Contents

Illustrations

Note: Figures 5-7 are reproduced from vols. IV and V of the *Scientific Papers* of Terada Torahiko, original edition, (Iwanami Shoten, 1936-38) by kind permission.

FOREWORD
Between Word and Number

The first time I went to a conference of literary scholars, a dozen years ago, I saw something I had never encountered at the physics meetings I was used to attending; the presenters read their papers from prepared texts, carefully moving, word by word, through their commentary and analysis. Scientific conferences were different. The scientists gave talks, of course, but words were only part of what they presented, and were invariably delivered spontaneously, without the discipline of a prepared text. Instead, the main weight of the presentations, the part that would be seen as persuasive or not, appeared on slides that conveyed scientific reasoning and theoretical or experimental results in the form of equations, tables, figures, and graphs.

These different ways of presenting ideas made me deeply aware of an aspect of the "Two Cultures" that remains insufficiently appreciated; while the broad range of subjects that today comes under the heading of literature takes The Word as central to what it does and how it conveys it, much of the equally broad range of subjects that constitutes modern science takes The Number—that is, quantitative reasoning—as a central mode of thought and expression.

It is a matter of common wisdom in our technological world that the humanities and the sciences must find ways to speak well to each other, and a matter of common concern that they rarely do. Their relation, or lack of it, has included mutual neglect, condescension, and outright hostility (evident in the Science Wars of the 1990s) on the one hand, and valiant attempts to bridge the gap on the other. But none of these efforts, be they blistering critiques or heart-felt desires to blend more fully, can be meaningful without mutual respect and tools for mutual understanding, which must include facility in the languages of The Word and The Number.

In *Metaphorical Circuit,* Joseph Murphy honors both science and literature by taking each seriously, as shown through his fluency in both languages. That creates an atmosphere in which progress can be made. Beyond this essential starting point, Murphy provides a path to deeper understanding as he examines a national culture where science and literature had reason to engage each differently than they do in the United States today. That *milieu* is the Japan of the early twentieth century, whose roots lay in the Meiji Period that began in 1868, when Japan, in Murphy's words, transformed itself into "a modern state with the technological capacity to defeat a Western power in the Russo-Japanese war and a GNP exceeding that of France."

In that place and time, Western science and literary thought were being rapidly grafted onto Japanese thinking, through a new national educational system that nevertheless drew on older traditions. The form of that legacy imposed a certain "symmetry" on the relation between science and literature that has not been naturally present in Europe and the United States. For that and other reasons, Murphy argues, early twentieth-century Japan offers a different and important perspective on how these two fields of knowledge may interact.

To bolster the argument, he relates case studies of Japanese figures whose work is interdisciplinary in one way or another. Two main examples come from the early twentieth century: the writer Natsume Sōseki (1867–1916), with his theory of literature that was expressed in quasi-mathematical form and drew upon cognitive science, and the physicist Terada Torahiko (1878–1935), known for his literary efforts as well as his science, and whose physics had a certain qualitative cast. Two others are contemporary literary critics: Maeda Ai (1932–1987) who in his *Literature in the City Space* used the abstract mathematics of topology to carry out literary analysis; and Karatani Kōjin, whose approach to literature also uses ideas from mathematics, as well as ideas of spatial representation, such as the notion of linear perspective.

These four figures represent two different ways to blend science and literature. Among them, Terada is the only one trained as a scientist. He began, however, as a pupil of Natsume Sōseki, who advised him to study physics. With this background, it is not surprising that Terada comes closest of the four to truly living in both areas, meaning he achieved success as both a research physicist and a literary essayist. While this may seem a modern version of Leonardo Da Vinci's life as scientist-artist, the reality is more complicated. As Murphy tells us, Terada might have received credit for a major discovery about how X-rays interact with crystalline solids, a stunning break-through at the time, except that he neglected to put his insight into mathematical form. Instead, the effect is associated with the Australian-born British physicist Lawrence Bragg, who in 1912–1913 summarized it in the equation known as Bragg's Law (Bragg and his physicist father were awarded the Nobel Prize in 1915 for their joint work on X-rays).

Terada's failure to use the language of The Number to convey his result in a scientifically convincing way is a cautionary tale, yet there is more to the story. Apart from his work on X-rays, Terada's research probed complicated physical systems that at the time lay beyond mathematical expression, and could be studied only qual-itatively. As Murphy rightfully points out, Terada can be seen as a proto-founder of today's new science of complexity. If his lack of complete focus on The Number cost Terada something in reputation within traditional physics, his ability to think qualitatively may well have opened him up to a new scientific approach as well as a literary career. The two kinds of thinking seem to have functioned in Terada's mind so as to give outcomes in both science and literature.

Natsume Sōseki, Maeda Ai, and Karatani Kōjin, however, do not operate in both areas; they bring scientific ideas to bear on literary questions, producing results written for other literary scholars but not necessarily for scientists. Each uses science or mathematics to produce a theory of literature—which has a fundamentally different meaning than a theory of gravitation, say. In physics, a theory is a set of mathematical relations that has two tests of value: does it accurately depict something already known about the physical world, preferably in broad terms that might include the entire universe? Can it predict something new about the world?

Both questions can be answered only by carrying out an exper-iment, and the importance of experimental verification in science cannot be overestimated. The equations of James Clerk Maxwell form a successful theory, because they describe all known facts about light waves, and unexpectedly predicted that electromagnetic radiation travels at the speed of light—both claims having been confirmed by experiment. But string theory, based on the concept that elementary particles are tiny vibrating strings rather than infinitesimal points, is in limbo as a valid scientific theory because it lacks experimental authentication.

It is typical of Murphy's rigor that just as he insists that The Number must have its place alongside The Word to properly deal with science, he explicitly points out that literature has no equivalent to the idea of verification by experiment. Hence there is an asymmetry in the meaning of "theory" in the two areas, a critical difference that yet seems to go unrecognized even by scholars working at the interface.

To me, the difference means that even if a literary theory includes scientific ideas, it should not be judged by the scientific criteria of experimental verifiability and predictive power. Rather the value of a literary theory is in providing a description of works of literature, their meaning, and their formative processes, to give a unified framework for their consideration and to stimulate further insights. In that light, the role of science in literary theory is to act as one more tool of insight or organization along with cultural and historical perspective, linguistic analysis, and so on. The association between the scientific ideas and the literary ones may be in broad terms, or at some level of detail (although a complete one-to-one correspondence between a scientific theory and a literary one appears impossible); but the important thing is that the appropriate use of science extends literary understanding.

One of Murphy's premises is the ready willingness of Japanese humanists to use scientific tools in this way. Thus Natsume Sōseki's major work *Theory of Literature* opens with mathematical symbolism in the statement "Let us represent the form of literary content by (F + f). 'F' here designates focal impressions, or concepts, while small 'f' designates accompanying emotions." Sōseki goes on to manipulate the symbols so as to represent different reactions a literary work may engender, and then brings in another set of scientific concepts as he relates the reader's response to contemporary models of human cognition.

In a similar vein, Karatani Kōjin's writings draw, for instance, on Kurt Gödel's famous Incompleteness Theorem, a significant achievement in twentieth century mathematics and logic. Maeda Ai did not hesitate to use the pure mathematical reasoning of topology, the study of geometric properties that remain unchanged when a figure is distorted, as a literary framework. But in Maeda's case, Murphy shows that it is not enough to use science; it must be used well. Maeda analyzes Flaubert's *Madame Bovary* in topological terms, and at first the story seems to perfectly match a topological model that differentiates among the inside, the outside, and the boundary of a figure in space; but as Murphy pushes the comparison harder—true to his program of serious consideration of both The Number and The

Word—we see the topological approach break down, leaving only an unconvincing literary application of scientific ideas.

Yet, as Murphy suggests and his own careful analyses show, the sheer readiness of literary scholars to engage science *as scientists know it* is far more useful than attempts to reduce science to a text like any other, a strategy that utterly denies value to the essential scientific properties of verifiability and predictive power. This is the issue that gave rise to the Science Wars. As its last chapter, *Metaphorical Circuit* incorporates lessons derived from the Japanese version of the science-literature interaction into an even-handed discussion of the origin and meaning of those Wars, and gives hope that the academic community might transcend them.

Although Murphy's examples show that science-literature interactions could be fruitful if our academic culture were to change, there remains a troubling asymmetry—the heavy preponderance of humanists over scientists in the interdisciplinary ranks. One reason can be found in Murphy's case studies, and in even a cursory look at other interdisciplinary work. While there are many examples that show the influence of scientific ideas on literary thought, it is hard to find examples where literary theory has affected how natural scientists think, and only marginally easier to find examples for social scientists. Until scientists perceive that literary approaches to science do not begin with the premise that science is just another text; until they believe that students of literature pay respect to words as both humanists and scientists use them, and to numbers as well, they will never believe that literature can illuminate science. Without that belief, in Murphy's words, the humanistic exploration of the relation between literature and science will continue to be "effectively carried out in the absence of the other discipline," without, I fear, much benefit to either field.

Sidney Perkowitz, Candler Professor of Physics
Emory University
Atlanta, Georgia
January 2003

Preface

Of the many interesting people to visit the University of Florida over the year I spent writing this book, two in particular stand out as impressive and legendary figures in their field. In a conference on Comics and Graphic Novels in February, 2002, Will Eisner, author of the seminal *A Contract with God* and father of the graphic novel form, expressed undisguised pleasure and satisfaction at the prospect of the storytelling form to which he had devoted his life finally being sanctioned and welcomed into the circle of the arts, the relief of a person whose long journey had come to an end. In an emotional keynote speech, Eisner charged the young academics present with the task of guarding and maintaining the standards that would distinguish good graphic novels from mere comics. In our tactful silence, we did not inform this great man that the reason he was here was that literary criticism had deconstructed such standards and paralyzed the faculty of judgement.

What one faced as a graduate student in the late 1980s and early 1990s was deconstruction, the stranger that came to town. I take deconstruction to be an essentially skeptical strategy, whereby one finds equally compelling arguments to hold both a proposition and its negation. It is inside, but it is also outside. It is one, but it partakes of many. It is A, but it is also not-A. It is poison, but it is also cure. This puts it in the lineage of the fierce Pyrrhonian skepticism of the ancient Greeks, as well as certain strands in Mahayana Buddhist thought, rather than the relatively mild modern doubts about error and the evidence of the senses. It is a powerful, radical strategy, not to be denied. It is as if literary criticism recognized Northrop Frye's charge that it lacked the poetics that would make it a discipline, but chose for its deductive principle one which made a hash of inductive practice. Such a strategy leaves scarce grounds for substantive knowledge claims, and brings practice to a halt. Of course, deconstructive principles were coupled with a variety of politically committed critical strategies. But there was always a doubt, as one read, that the radical practice resulted not from theory but from ethical decisions that were extraneous to theory.

This is not to make the claim that deconstruction is a fashion. It is rather a religion. The second seminal figure to visit here at the University of Florida was of course Derrida, who in a series of appearances in March of 2001 seemed ironically aware of this status, characterizing himself as a preacher and asking his listeners to make a leap of "faith" prior to beginning his discussion of issues such as lying and the death penalty.[1] The inversion of doubt and faith is intrinsic to religion. One believes, then one understands. Zizek (1989), following Anselm, recognized in this a problem common to the institutions of Christianity and Marxism. Arguments based on faith are instantly persuasive to the assembled congregation. But how to convince someone outside the community of the faithful. There is no exterior to the community of the faithful, one can hear the choir intone. The failure of academic cultural critics to make even a dent in the public discourse on the second war with Iraq, however, testifies to the acuteness of this problem.

The humanities imagines that if they are consumed by doubt, legitimate doubt coming from a serious philosophical position, so too must the sciences. But that did not happen. How to account for this difference? Steady reading of any of the lead professional journals in the natural sciences will show them to be in an unprecedented state of robust health and confidence. A reductive program of knowledge has closed down one by one domains formerly thought not to be amenable to physical description, and gaps in the physical explanation of the world, once covered by terms like spirit, soul, life principle or nervous principle, have fallen one by one, such that a closed and complete account of the world in terms of physical processes seems possible in principle, even if the details elude researchers now. This reach of the sciences extends to the causal, physical, explanation of mental phenomena, in principle literature as well, in terms of processes in the brain, a program being pursued outside humanistic territory by interdisciplinary teams of biologists, neuroscientists, electrical engineers and specialists in artificial intelligence. If anomalies appear at the outer extremes of physics, these facts are taken as brute. Meanwhile, engineers use this program to continue modestly changing the world.

---

1. This sort of combination was reprised in 2003 with Bill Griffith, author of the *Zippy the Pinhead* comics, followed by Julia Kristeva.

The inexplicable dominance in the humanities of hundred year old linguistic models based on Saussure, and a pscyhology based on Freud has made it difficult for literary critics to recognize or account for these changes in the other disciplines that study language, perception, the mind and consciousness. As a result, literary criticism has operated since the 1970s with an image of science that is a caricature, or a mirror image with which it imagines itself having a conversation.

The possibility of communication is what this book is about. In Japan, I found there has never been the luxury of generating science as a fantasy object. The stakes in recognizing the technical extensions enabled by science were too high, and science was present in the literary curriculum. This engendered to my mind a remarkable mutual recognition.

The basic ideas for this book were formulated in a graduate seminar paper in 1992 that forms the basis for chapter 5. This was two years before Gross and Levitt's *Higher Superstition*, and prefigures their argument substantially. Intervening was a period of five years where I learned my craft. In this time I wrote a dissertation on "degraded" forms of Izumi Kyōka, and published an unrelated series of five articles on interactions between Japanese film and literature. I consider this honest work, but my concerns were elsewhere.

My concerns lay in reconciling my experience in engineering and my experience in the humanities, because both are part of my identity. I value the method of probing of humanistic inquiry, specifically literary theory. That is why I hungrily took humanities electives while acquiring my first education in Mechanical Engineering, and that is why I switched into the humanities for my graduate work. Literary criticism still asks the big questions, the kind that occur to one as a child or sophomore not because they are simplistic but because they are fundamental, and for which adults have not found an answer. Coming from engineering, though, I know that engineers think about these kinds of thing as well. They will not be browbeaten however, into accepting half-truths that skew their concerns about the physical world.

My dialogue during that five years was primarily with the Japanese thinkers I met in books. A panel I participated in at the Society for Literature and Science Meeting in 1998 first gave me the idea that this problem might be formulated in a way that was of general interest, and might admit of study. I would like to thank Laura Otis and Sidney Perkowitz for organizing this panel, and providing me with continued friendship and inspiration to pursue the problem. I would also like to thank an unnamed colleague at the University of Florida, who dropped by my office one evening the same year as I was dreamily engrossed in an engineering text on efforts to model vision with VLSI circuits, and asked rather skeptically what this had to do with my work in literature. An excellent question. I see these two points as moments where a project snapped into place in my mind, and revealed itself as workable. I think I was aiming in a clumsy way at starting some kind of dialogue. That dialogue I hope will begin right here, between the Foreword by Sidney Perkowitz and the body of the book.

I intend these essays to be enjoyable. For the attentive reader, I hope there will be unexpected turns as I follow these thinkers over this broken field. For the inattentive reader too, there will be plenty to react to. Beyond the pleasures of argument, though, and what William Haver has called the "labor of refusal" called forth by polemic, I hope to have brought up some original points. I believe the connection between Sōseki's *London Tower* and *Theory of Literature* to be new, as with the reading of the missing fourth possibility in {F+f}, and the connection between Terada Torahiko and D'Arcy Thompson as progenitors of complexity, arrived at in a speculative way, but confirmed in the citation by Thompson of Terada's pupils in the 2nd Edition of *Growth and Form*. But there is also implicit in my discussion of literature and science a refusal or a surrender of literary criticism, my chosen profession, as a *serious* undertaking, that is to say, one that produces substantive knowledge claims. Hence my next work will explore the idea of play and excess outside of an instrumental economy, or a particular capitalist economy at least. I am interested in working hard without being useful to projects I despise, in wasted efforts without return.

An early version of chapter 2 was published in *New Critical Perspectives on Twentieth Century Japanese Thought*, Monnet, Livia, ed., University of Montreal Press, 2001, and is reprinted with kind permission. I've incurred many debts. At Cornell, Brett deBary, Naoki Sakai, David Bathrick, Susan Buck-Morss, Ayako Kano, Mark Anderson, Mike Bourdaghs, Susan B. Klein, JoAnn Izbicki, Mark Driscoll, Robert Steen, Guy Yasko, and Michele Li contributed to a remarkable graduate program that was generous, intense and highly supportive. Professor Iyotani Toshio of the Tokyo University of Foreign Studies provided an intellectual home while researching in Japan, and Professor Yamanouchi Yasushi, Narita Ryuichi, Oguma Eiji, Iwasaki Minoru, Professor Kamei Hideo and Karatani Kōjin all provided guidance and support. Komori Yōichi kindly allowed me to sit in on his Sōseki seminar at Tokyo University in 1994, where I first learned to see Sōseki as a theorist. Livia Monnet provided key support to me in the early years, along with Tom Lamarre, Charles Inouye, and Margherita Long. Johlyn Dale reminded me at a crucial time of the romance of writing. Norma Field and William Sibley provided generous hospitality and mentorship during a postdoctoral fellowship at the University of Chicago from 1997-98. This study would not have been possible without the year provided me there. The Asian Studies Program and the Humanities Enhancement Fund at the University of Florida provided material support at crucial times for research and travel. I would like to thank Karen K. Smith for editorial assistance, and the anonymous readers at CEAS for suggestions that directly improved this manuscript.

I owe a great debt to my colleagues in AALL at the University of Florida, in particular to Avraham Balaban for a steady hand, and Ann Wehmeyer, Yumiko Hulvey, and Susan Kubota for close support and friendship. Fred Gregory and the Science Studies Reading Group, Harry Paul, Robert D'Amico, Galia Hatav, Scott Nygren, Maureen Turim, Luise White, Hunt Davis, and Norman Holland provided connections across field, while Professor Kenneth Heilman and the CNS Seminar in the Neurology Department allowed me to observe the really big differences. I would like to particularly acknowledge the superb undergraduates in my JPT 3120 course in 1998 and 2000, and JPT 3500 in 1999, who demonstrated fearless curiosity and willingness to explore, and in whose lectures a good bit of this was worked out. I'd like to thank the night manager at the Winn-Dixie on 16th Street for expert provisioning, my sons, for never failing to amuse, and my parents, Joseph and Dolores Murphy for never failing to support. This book is dedicated to my wife. All these people helped me, none is responsible for the failures, the mistakes and the work still to do.

Joseph Murphy
Gainesville, Florida
January, 2003

1 Introduction

Purpose and Scope of Study

This study explores the negotiation of a certain boundary in the production of knowledge by thinkers in early twentieth-century Japan. These include novelists Natsume Sōseki, Mori Ogai and Edogawa Ranpo, physicist Terada Torahiko, and contemporary academic critics Maeda Ai and Karatani Kōjin as they write about this period. This boundary is persistently at issue in North American intellectual discourse as well. It is the boundary between literature and science.

The terms "literature" and "science" have shifting and capacious meanings, and so I should offer working definitions. I will allow the sense of literature in this study to shift and change depending on the concerns and context of the speaker. Because the ultimate stake of this book is in reconceiving the term. However literature will tend to stand in for the humanities in a more general sense as the form of understanding most clearly opposite the sciences in the division of knowledge that characterizes the modern university. Borderline disciplines become social sciences insofar as they borrow the rhetoric and methods of the sciences, and they become humanities insofar as their approach is literary. By science I have in mind an operational definition wherein the defining feature is that the test of all knowledge is experiment.[1] This definition has the advantage of making no claim about what happens in the stage of experiment, and stresses instrumentality,

---

1. This understanding is drawn from reading in the work of physicists, principally Percy Bridgman, Werner Heisenberg and Richard Feynman, discussions with reader-response critic Norman Holland, and the critical realist philosophy of Roy T. Bhaskar. It is open to revision.

1

that is to say, the ability of scientific knowledge to enable a direct, practical intervention in the material world, rather than truth as the criterion of scientific knowledge. However it is true that scientific knowledge is instrumental. The natural sciences tests its knowledge through experiment, and the claim of the human sciences to be science rests on this. There is no corresponding concept in the humanities.

The idea of modeling, in particular mathematical and statistical, has been found to be crucial in this separation of the field of knowledge. It is the contention of this book that efforts in the humanities over the last twenty years in North America to comprehend the sciences, both in the sense of understand and subsume as a part of itself, have been hampered by a desire to remove mathematical expression from the equation. This has resulted in a certain tone-deafness to humanities-based discussions of science, which manifests itself as a defensive insularity or in the conceit that science is just another narrative, text or discourse, which allows the humanist to discuss the fascinating themes of science, and even partake of its rhetoric of rigor, without encountering the heterogeneity of the pursuit of knowledge in the sciences. And it is this heterogeneous quality in the field of knowledge that is my concern.

In the context of a set of responses to efforts by experimental psychologist Herbert Simon to outline a cognitive scientific approach to literary criticism, Norman Holland applies Ockham's Razor to the problem of the humanist's reaction when faced with real science:[2]

> I wish my colleagues in criticism and theory would accept Herbert Simon's offer of cognitive science. He suspects they won't and wonders why. Why are our theorists so reluctant to accept a view backed by a vast psychological research into perception, remembering, knowing, and reading?
>
> Most humanists were good in school at English and bad at math. They fear science. For support from another discipline, they prefer philosophy to experimental psychology. (Holland 1994, 65-66)

Indeed, it is a peculiarity of higher education in the United States that it is possible to advance to the Ph.D. level in the humanities without encountering a course in calculus, a basis of representation as fundamental to the modern physical sciences as French or English

---

2. That is to say, as it circulates among the community of specialists that produces it, as opposed to secondary and tertiary summaries.

is to Comparative Literature. Accepting for a moment the rhetorical setup, the question then arises of why a group of scholars who are for the most part bad at math and fear science would have a strong desire to talk about science. The answer is interdisciplinarity. However, the posing of problems like "literature and science" by scholars in a humanities so configured risks a curious kind of interdisciplinarity, one carried out in the absence of the other discipline. This has made it easy for opponents to dismiss the humanities-based critique of science in North America as incompetent, and given the critique itself an anxious misrecognition of its status vis-a-vis the "better endowed" institution of science.

This book takes the form of a series of case studies of individual writers and thinkers, but also advances an overall argument. The overall argument of this book is that the theoretically-informed humanities is going down the wrong path in its dealings with science. The near universal gesture of science studies since the 1970s has been to try to erase the difference of scientific knowledge. Cover terms include metaphor, text, discourse, and narrative, however the idea that the social construction of knowledge implies that science can be comprehended by the humanities through its own formalist or semiotic procedures constitutes now common sense for the field, able to be introduced without argument. It is against this common sense that my study is aimed. In this I concede the initial point of the recent Science Wars, that the humanities cannot comprehend science without venturing off the safety of humanistic terrain and onto the disconcerting heterogeneity of interdisciplinary ground.

In attempting to make a metaphysical object out of science, hence an object which can be handled ironically with the standard critiques of metaphysics, the humanities is caught in self-reflection. Against this synthesizing tendency, I locate a series of twentieth-century Japanese thinkers who risked this interdisciplinarity and discovered instead an irreducible heterogeneity to the field of knowledge, that cannot be bridged. In going outside the context of North American debate to the seemingly faraway land of early twentieth-century Japan, I am looking for cases that satisfy two conditions: One, that the movement between literature and science was a central intellectual tension for the writer involved, and two, that they keep the question of mathematical representation in the equation. That is to say, I am interested as a cue in points where mathematical representation, not literary representation, emerges as an issue in negotiating the boundary between literature and science. Such points are not difficult to find.

From Natsume Sōseki's formula-riven *Theory of Literature* (1907) and modeling of civilization as a thermodynamical system, to the set theoretical premises informing Kyōto School philosophy,[3] to the technical issues undergirding Karatani Kōjin's reformulation of institutionalized deconstruction, to the use by Ozawa Masachi of George Spencer Brown's *Laws of Form* in modeling an open-ended social practice, intellectuals in Japan have a habit of bringing mathematics to their encounter with cultural problems of modernity. In this I hope to locate that peculiar hybridity that marks truly interdisciplinary work, the sense that pieces are being brought together that do not really fit.

In this I run the risk of repeating a move that is familiar in post-orientalist scholarship, that of locating in Japan a "pure-land across the sea" where the dualisms and ills that beset modern Western culture are found to be non-existent or attenuated.[4] There are two points which complicate this: First, I am arguing for a dualism in Japan. Versus a utopian desire for coherence and unity in the field of knowledge in contemporary North America, Japanese thinkers recognized its fragmentation as foundational. Secondly, I argue for a causal explanation in terms of specific institutional grounds. Japanese intellectual space from roughly 1890 to 1945 is unique in that the disciplinary divisions of the modern university instituted in the span of a generation along contemporary European lines were inhabited by scholars who retained a corporate/corporeal memory of intellectual work under a different regime of knowledge in the form of Chinese studies. These scholars neither feared the formalization involved in math nor took for granted the immiscibility of knowledge produced by the humanities and the sciences. It is possible to argue inductively that this

---

3. This would include both a set theoretical logic in Nishida's sounding of the problem of general and particular, Tanabe Hajime's logic of species and genus, and the general conception of the Greater East Asia Co-Prosperity Sphere as the general site where all sorts of particulars can coexist, as well as partaking of the particular concerns of nineteenth-century mathematics, such as the handling of infinity and zero, discontinuity, self-reference, etc. Naoki Sakai first directed my attention to this aspect of Japanese thought in a seminar in the early 1990s. See his elucidation of the intellectual underpinning of Tanabe's logic of the species in Sakai (2000), especially 521-522, n. 23.

4. Lehman (1987) points out the paradigmatic status of this move in film scholarship, while Kitagawa's location of a "unitary world of meaning" in the religious consciousness of archaic Japan, whatever its status as anthropology is also a point for point resolution of Hegel's notion of eight "dualities" that beset modern man (Kitagawa 1987; Forster 1998, 17-44).

space oriented scholars toward interdisciplinary work, and toward formalization when addressing humanistic problems of modernity. This makes Japanese intellectual history a good site to sound the relation of literature and science. Indeed, beyond the relevance for contemporary debate about the sciences, one might discern an interesting new aspect of Japanese artistic production if one brings an eye attentive to math and science issues.

Behind the consistent tone-deafness of humanities scholars to science issues is a deeper, institutional problem with science literacy. James Paxson has called calculus the "price of admission" to the discourse of the sciences, while Asada Akira calls it the "absolute, minimal common sense of modernity" (Paxson 1997; Asada and Fukuda 1999, 143), and it should be as incomprehensible for a Ph.D. candidate in science studies to be without calculus or differential equations as for a Ph.D. in Comparative Literature to be without French. The near evacuation of mathematics from the curriculum of general education in the humanities, by contrast, is arguably a primary cause of the sterility of the debate over literature and science in the United States.

Hence, though this book is concerned with science and literature as modes of understanding structuring the field of knowledge, behind it will always be the question of mathematical representation. This is not a concern with the technical elaboration of specific types of mathematics, nor does it pretend to deal with the glamorous upper reaches of modern math—with the higher levels of mathematical logic and number theory, the algebra of sets and number fields, projective and non-Euclidean geometries, which would constitute a species of exoticism at this point. It is rather a concern with the introductory levels of mathematical expression and their relation to processes or states of affairs in the physical world as a threshold over which Japanese thinkers pass, or don't, as the case may be, in moving between science and literature. It is at this point that math acts as the grammar and syntax of scientific expression, and I would argue that it is a feel for the way the concepts and domains of mathematics are mapped onto the physical world at this level, and not an ability to converse in any particular mathematics, that constitutes the basis of science literacy.

One might well think that my intention, by putting math "back into the equation," is to build a bridge between literature and the sciences, to produce policy-relevant recommendations about school curricula, and in general heal the rift and facilitate a climate where the two domains of knowledge can come together and communicate.

However, what I find in thinkers of this period is that a concern with literature and science brings them repeatedly to a boundary, beyond which nothing can be said. In this sense, this study ends not in curing a rift between literature and science, but in affirming and securing a difference between scientific knowledge and other more "metaphysical" kinds of knowledge like literary criticism. Because workers in the rarefied environment of the Imperial Universities took seriously the question of interdisciplinarity, and brought to the task a sensitivity to mathematical representation, Japanese thinkers in this period provide a framework that can fully acknowledge the power, precision, and instrumental validity of scientific knowledge. However, the clear recognition on the part of Japanese thinkers of the instrumental role science plays in the economic, technological and military adventure that is the West allowed them to question whether this grants science any privilege in an absolute sense, and to discern a domain of problems that cannot be posed scientifically. That these realms of reality that remain untouched by scientific explanation are therefore the purview of the discipline of literary criticism as presently conceived, unfortunately, does not at all follow. To borrow the words of Kant, this critique is negative in that it seeks to forbid a certain facile way of imagining that literary theory has something meaningful to say about science. However this move is positive in that it secures a space for faith, which in this secular, digital millennium seems to be all that keeps the enterprise of literature going.

Why Early Twentieth-Century Japan?

Because this study is framed in the field of literary criticism, rather than history or social science, its starting point must be the record of the anxieties of intellectuals, and in this I have perceived three things: First, the relation between literature and science seems to be a central animating tension for intellectuals in Japan throughout the twentieth century. Second, scientists as well as humanities scholars are involved in the discussion, and third, thinkers seem to presuppose a level of math and science literacy that indicates a real interdisciplinary concern, not a dilettantish curiosity. These points stand out in the greatest contrast to the painful mutual incomprehension of similar discussion in the contemporary North American academy.

It is the contention of this study that early twentieth-century Japan is a *better* place to think about these rifts and heterogeneous features in the intellectual field than the contemporary United States, for all the sophistication of the critical scene in regard to science and technology studies. Though the above claim has to emerge from the chapters to follow, a couple preliminary indications at least raise the possibility of a difference in the Japanese intellectual milieu pertinent to this problem: First is the odd prominence of literature in major histories of modern thought, indicating a different status for literature in the intellectual field. For example, the philosopher Kōsaka Masaaki's monumental *Japanese Thought in the Meiji Era* (1958) includes an 80 page section on Natsume Sōseki and Mori Ogai, who are novelists, and accords the relatively short-lived and aesthetically mixed Japanese Naturalist movement a central role in instituting a modern scientific sensibility in Japan. Social scientist Maruyama Masao's highly influential *Nihon shisōshi* (History of Japanese Thought, 1961) devotes nearly half its length to reworking an opposition between "literary" and "scientific" modes of thought from prewar Marxist debate as key determinants in the absorption by Japanese of external ideas. Second is the manifest investment by scientists in the practice of literature. Mori Ogai, arguably the most powerful figure in Japanese literary history, was famously a practicing physician and surgeon general of the Imperial Army, while many well-known lesser figures, such as Terada Torahiko, critic Kato Shuichi and novelists Kita Morio and Abe Kobo were formally trained in science or medicine. Ishihara Jun, the relativity theorist who studied under Einstein and Sommerfield, and escorted Einstein during his 10 week visit to Japan in 1922, quit his post as a professor of physics to pursue a career in poetry. It is further not at all unusual in Japan for working scientists to publish books of literary criticism or biography.[5]

There are a number of reasons to think that the relative prominence of literature as a problem for intellectual historians and working scientists in Japan, that is to say the appearance of the problem outside academic departments of literature, indicates a serious difference between Japan and North America in the broader institutional conditions relating to intellectual life. These must be sought in the institution of a modern university in Japan in a single generation as part of the broader package of reform that defined the Meiji state.

---

5. Studies I have used in preparing this book include *Miyazawa Kenji—Shijigenron no tenkai* (1991) by physicist Saito Bunichi; *Terada Torahiko—Sono sekai to ningenzō* (1971) by Ota Bumpei, then Computer Operations Director for Hitachi; and *Sōseki ga mita butsurigaku*, v. 1053, Chukō Shinsho (1991), by Koyama Keita who holds a doctorate in the physical sciences.

The story of the transformation in the Meiji Period (1868-1912) of the heterogeneous social and political formations of the Tokugawa period (1603-1868) into a modern state with the technological capacity to defeat a Western power in the Russo-Japanese war and a GNP exceeding that of France is well known.[6] From political parties and prisons, to legal institutions, post and rail, joint-stock companies and the modern novel, a spectrum of institutions deemed necessary to put the nation of Japan in a position of parity with western powers were assembled by its leaders from carefully studied models in Europe and the United States and put in place in the three decades between the Meiji Restoration in 1868 and the renegotiation of the unequal treaties with England in 1894. This program transformed the productive activities of the people, if not their sensibilities and daily lives, and caught the elite in a competitive program of perpetual incorporation and change.

As part of this broad transformation, the institution of universal, compulsory schooling at the elementary level was relatively straightforward in concept, and occurred within four years of the Meiji Restoration in 1872. However, it took somewhat longer to tame the variegated system of schooling of the Tokugawa period, with its temple schools and merchant and domain academies, its assorted curricula and wide disparities in educational attainment by class, gender and region, into a rationalized system of higher education. The process begins with the establishment of the Kaisei Gakkō in 1870, a rough, chaotic predecessor to the Imperial University whose feeder schools were largely former domain academies, and which lumped together boys from 12 to 25 with diverse backgrounds and attainments, rushing prodigies through and barely containing the fractious older boys overwhelmingly of samurai ancestry. Though the school was renamed the Imperial University with the addition of a faculty of Medicine in 1877, it would retain the rough heterogeneity of the Tokugawa legacy for much of the next decade. It would be in the 1880s that the modern university took shape, with the addition of faculties of Law, Engineering and Agriculture, the systematization between 1886 and 1890 of Higher Schools One through Five as regional feeder schools, and the segregation of the elite students in the resulting five year, boarding school curriculum. This resulted in a modern university with a fully

---

6. The story is best told by the Japanese themselves. See Okuma, ed., *Fifty Years of New Japan*, 2 vols. (1909), published originally in English.

instituted disciplinary division of labor on German, English and American models by 1890 (Roden 1980; Bartholomew 1989).

John Guillory argues that the American university also took its present form over the same period in the late nineteenth and early twentieth century (Guillory 1993). However, the United States had essentially transplanted a European system of colleges and seminaries during its colonial period, hence the transition to a modern university had the same character of imperceptible change and gradual professionalization of existing disciplines. The Japanese case is unique in that the transition effected in a single generation is from a wholly different system.

The question of form is most speculative and refers to Sasaki Chikara's argument that the term *kagaku* (科学) carries in Japanese both the sense of "science" and the sense "division into faculties," i.e., the sense of knowledge in general (Sasaki 1996, 15-18). The prefix *ka-* in *kagaku*, also appears as a suffix in the division of knowledge production into faculties, as in the Faculty of Letters (*bunka*), Law (*hōka*), Business (*shōka*), Agriculture (*nōka*), Engineering (*kōka*) and Natural Sciences (*rika*). Schematically:

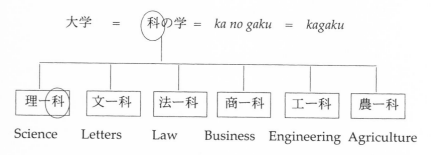

Fig. 1. Division into Faculties of the Imperial University

This does not inhere in the suffix "sciences" in English, as in social sciences, natural sciences, life sciences, human sciences, as the addition of the suffix always indicates the subjection of a domain to the methodologies and protocols of the natural sciences, and retains its opposition to "literary" or "humanistic." In the inclusion of science as one of the faculties in *ka no gaku*, there is a self-referentiality that veers close to the formalist paradox of the set that contains itself as one of its members. This self-referentiality at the institutional level of one of the terms means that, despite the use of the European model, the pairing of *"bungaku vs. kagaku"* does not call up the same clear opposition as "literature vs. science," running rather into the rhizomatic structure of *ka no gaku*. In this it is possible that occupants of the faculties of science and letters might not find it logically necessary to repudiate the position of the other "root and branch" in order to keep secure their own position.

The question of the rapidity of the institution is more crucial, and points to the coexistence of two regimes of knowledge within the modern university in Japan, namely a pre-Meiji legacy of Tokugawa period medicine and Chinese studies, in which gentlemen scholars pursued the study of nature as part of a broader program of cultivation and in which the opposition between literature and science would appear as a symmetrical difference between domains for the investigation into truth, and the modern university encountered and instituted in the 1880s, in whose disciplinary structure science retains its general sense while literature is marginalized as the specialized concern of a single department. I would argue that this coexistence of two regimes in the Imperial University, embodied in the individual educational histories of the generation that came of age in the 1880s, rendered the asymmetrical relation between literature and science "strange" for its occupants.

A third reason why it makes sense to find a more interesting relation between literature and science in early twentieth-century Japan relates to script reform. Math is a mode of expression with relatively tight rules and patterns of transformation and a domain in a variety of number systems. Science maps these domains and patterns to the physical world, and is enabled to speculate. It is a mode of expression the American humanities does not speak. It seems odd to use math to work out humanistic questions, estranging, even damaging to the understanding allowed by literature. But writers and intellectuals in early twentieth-century Japan often *did* slip incongruously into this mode of expression, and it is the thesis of this study that attention to these points when they jump out of the literary text, leads one systematically to a larger problem—the relation between literature and science. For intellectuals in this period, the Japanese language which seems now the unproblematic locus of expression and identity for Japanese literature had a similar estranging quality, both in that the locus of instruction for many in youth was Chinese studies, and in that this Japanese itself was in the process of being formed and standardized by the state up until 1945. And one encounters in the use of language in literature the same experimentation, the same working out of possibilities through trial and error, and comparison with experience that one associates with science. That is why *naturalism* is so central for Japanese literary history. Faced with the fascination with punctuation, the incongruous slippage to numbers and other forms of symbolic notation in late nineteenth-century writers like Tsubouchi Shōyō, Ozaki Kōyō and Yamada Bimyō prior to standardization, it is possible to speculate that expression in the writing system was *just as alien* for Meiji intellectuals as expression in numbers. It was not a matter of writing in Japanese being natural and transparent, and symbolic representation jarring, suggesting a further symmetry.

The role of Chinese studies as a locus of identity for Meiji intellectuals is elaborated in detail by Maeda Ai in *The Formation of the Modern Reader* (Kindai dokusha no seiritsu, 1989) and by Kato Shuichi in volume three of his *A History of Japanese Literature*. In tracing the formation of a modern literary consciousness in the nineteenth and twentieth centuries, both Maeda and Kato distinguish generations depending on how their education fell out in relation to the changes of the Meiji period. In contrast to the generation of Restoration luminaries such as Fukuzawa Yukichi (1835-1901) and Nakae Chōmin (1847-1901), whose intellectual formation occurred entirely before the Restoration, the Age of Meiji belonged to the peculiar "generation of 1868," whose hybrid educational formation marked them as unique, and never to be repeated:

> The writers and thinkers who were born in the twenty years around the Restoration, from 1860 to 1880, were educated by this new university system, an experience that sets them apart from the generation of Fukuzawa Yukichi and Nakae Chōmin. However, the primary education that they received was very much in the Tokugawa mould, being largely composed of the study of Chinese classics. (Kato 1979, 107-108)

Chinese studies (*kangaku*) does not refer here to Confucianism (*jugaku*) as an ethical system, but to the broad range of cultivated learning, including rhetoric, ethics, natural philosophy, metaphysics, poetry and music absorbed through the recitation and study of the Chinese classics from childhood. That the default program of cultivation in the Tokugawa period for the ruling *bushi* (samurai) class, as well as the emerging farming and merchant elite was Chinese studies is not a point that is typically foregrounded in English language area studies, designed to pick up what is Japanese in the modern tradition. The problem is that this relation to Chinese studies was not just a supplement, but formed rather the locus of identification for the elite against the claims to universality of the West in the arena of reflection, and the idea of reference to an exclusively Japanese aesthetic and cultural tradition in order to secure identity does not gain wide currency until the 1910s with the dissemination of the ideas of Okakura Tenshin, Ernest Fenollosa and Watsuji Tetsurō, a move identified by Japanese critics specifically as a form of Romanticism, in close complicity with Western modernism's appropriation of a *Japonist* aesthetic (Karatani 1997; Karatani 2001; Isozaki 1999). The legacy of this for Japanese intellectuals is reflected in Kato's own demanding work, a point that has to be explained for the reader in English:

> What Mr. Kato has done here is to choose to define literature not as other historians of modern Japanese literature have tended to do, i.e., as fiction, poetry, drama, and essays in the lyrical vein, which would have made his task easier, but as writings of any category that played an important role in the development of modern Japanese thought. . . . The reader who expects the book to be mostly about novelists (they occupy most of the space in other accounts of modern Japanese literature) will be disappointed.[7]

That disappointment is the product of a very specific expectation, conditioned by the particular and historically contingent division of knowledge in the modern university. In *Origins of Modern Japanese Literature*, Karatani Kōjin argues that a clash of sensibilities in the Meiji period gave the generation of Natsume Sōseki and Mori Ogai a perpetual distance from the forms of modern literature they deployed so well. Though according to Karatani these modern "constellations" appear as self-evident to a Japanese reader as they do to us today, observation of the peculiar context of late Meiji Japan renders seemingly natural categories like landscape, sickness, the child, senses of perspectival depth and psychological interiority, through a process of blindness and insight, visible as constructions. Karatani confirms the importance of "speed" in this process of making strange:

> I'd say the institutionalization of theory, whether it be deconstruction or whatever, is virtually complete. That's why its important to go back to the starting point, and further, I think the Japanese context is an extremely auspicious place to do it. When

---

7. Edwin McClellan, "Foreword" to Kato (1979), ix-xi.

I wrote *The Origins of Modern Japanese Literature* that's what I had in mind. When people say "origins" they are usually pointing to the West, and even there to the distant past. However in a certain stretch of time in Japan's Meiji period, a number of things that, because they occurred over the course of several centuries in the West have rather been rendered invisible, appear in a highly condensed form. By taking a close look at these isn't it possible to call into question things that have become self-evident in the West. (Karatani, Comer, and Tsuge 1991, 20)

Barbara Herrnstein-Smith identifies as a defining feature of contemporary intellectual life in the United States an adversarial dynamic she calls "asymmetry," using the recent Science Wars as a chief exhibit. Asymmetry implies an epistemic self-privileging whereby, despite there being legitimate issues at stake in intellectual controversy, one must repudiate the other's position at the root level, denying any validity whatsoever to their claims, and attributing differences not to "more or less extensive differences of individual temperament and intellectual history, as played out within more or less different disciplinary cultures and sustained under more or less different epistemic conditions" (Herrnstein-Smith 1997, 136-137), but to benightedness or a deficiency in the other side. Hence discussions are cast as a War. The form and the rapidity with which the modern university was instituted in Japan made thinkers aware of the contingency of their situation, and give reason to think that debate over the relation of literature and science might take a different shape.

Why Literature and Science?

In the introduction to his book, *Hegel and Skepticism*, Michael N. Forster discusses the often unintended comic potential of juxtapositions across disciplines or areas of specialization:

[B]ooks of the "X and Y" variety ("Wittgenstein and Buddhism") often share with comic partnerships of that variety ("Laurel and Hardy") a capacity for eliciting amusement by their juxtaposition of incommensurable quantities. This is a virtue in a comic partnership, but not in a piece of philosophical literature. (Forster 1989, 1)

The juxtaposition of "Literature and Science" appears to be another example of such a comic partnership. Yet, interest in the juxtaposition does not fade. There is a Society for Literature and Science designed to effect the partnership, journals devoted to it, international conferences on the topic, even a volume claiming that the pairing constitutes a "Third Culture" in the modern university (Shaffer 1998). Werner Heisenberg hints at a deep connection between the two fields in observing that "Newton devoted a large part of his life to philosophical and religious investigations and it is probably correct to say that the world of poetry has been familiar to all really great scientists" (Heisenberg 1949, 67). However, in the asymmetrical character of intellectual life in the United States, the idea of "poetry for physicists" takes on a different meaning, ensuring that this type of discussion is overwhelmingly the concern of the side of literature, with little participation or interest by scientists.[8] Indeed, I argue in chapter 6 that dissatisfaction with the heavily "literary" state of this inquiry was the impetus behind the debacle of the Science Wars.

Yet, the juxtaposition is not as absurd as it first seems. Though literature is now typically understood in a narrow sense as "fictional texts of a certain quality," and confined to certain departments, the term has an older sense, traced by Raymond Williams in the West up to the eighteenth century where it stood for letters or cultivated learning in general (Williams 1976, 183-187). Under this guise, a general opposition between the study of nature and the rational and speculative operations of the mind appears in many systems of thought, in the opposition of Physics and Metaphysics in the Greeks, of Rhetoric and Investigation of the Principles of Things in the Chinese Learning (*kangaku*), and in the opposition of Natural Philosophy and Rhetoric or the Literary Arts in Renaissance Europe, and one may recall in this context that Newton's *Principia* laid the mathematical foundations not for "natural science," but for natural philosophy (Obayashi and Morita 1994, 3-7).

Any formation of this sort allows a certain symmetry to the juxtaposition of literature and science, and the comic dissymmetry of con-

---

8. A glance at the biographies of contributors to any of the last 8 issues of *Configurations* (quarterly journal sponsored by the Society for Literature and Science) or the departmental affiliations of participants in the annual SLS convention will quickly confirm this. As Jay Labinger, of the Department of Chemistry, Cal Tech put it in his statement for candidacy for president of SLS, "I joined SLS in search of ways to bridge the gap between the "two cultures"—a gap which I believe arises much less from hostility (as the current emphasis on science wars would have it) than indifference—and found an organization which has done a great job of providing an environment in which a broad range of scholarly activity can (usually) comfortably coexist. With regard to the more difficult problems of promoting serious interdisciplinary dialogue, understanding, respect, and cooperation, though, I feel we still have a good way to go" (Society for Literature and Science, 1998 Candidate Statements and Ballot).

temporary debate, then, can be seen to derive from the local and historically contingent nature of the disciplinary structure of the modern university, where literature has been restricted to a narrow subfield, while science retains its general sense of the systematic inquiry into natural phenomena. Under such a regime, efforts to link literature and science typically involve discussion of the thematic citation of science in literary works, or the location of the operation of metaphor and other figurative language in the putatively transparent language of scientific exposition. The first is trivial as an interdisciplinary point because it tends to neglect the role of mathematical representation in defining and exploring scientific concepts, hence to rest on vague analogies with words like chaos or catastrophe, while the second is problematic because scientists use metaphor explicitly and consciously in the formulation and public communication of their ideas, and seem perfectly comfortable discussing its implications and place in the scientific process.

Indeed, the fierce anxieties on display in the Science Wars at the end of the 1990s only make sense in the context of the movement over the past decades by the discipline of literary studies to reclaim its former generality under the guise of literary theory. In this sense, the pairing of "Literature and Science" is not comic, but can be understood to form a set with other pairs like "Science and Religion" or "Science and Philosophy" as an effort to delineate a domain of truth and intellectual activity for the others of science in which the instrumental claims of scientific knowledge would not carry a privilege. Here religion would delineate a mystical or spiritual domain, and philosophy and literature figure domains of humanistic inquiry based, in the case of North American discourse, on analytical and continental hermeneutic traditions of philosophy respectively. Yet, given the impasse evident in the Science Wars, it would be difficult to claim that this partnership has been effected to the satisfaction of either party.

Concern about the separation of "Two Cultures," and wars about the relation of literature and science can be understood to be in a causal relation to a basic problem: a split in the modern university between the sciences and the humanities that did not obtain in prior formations of knowledge. If this is the case, what is unique about Japan in the period under consideration is the coexistence of a fully instituted university system, along with the legacy of broader conceptions of the division of knowledge under the heterogeneous formation of knowledge in the Tokugawa period. That is to say, here is an intellectual milieu that embodied *both* conceptions of the opposition between literature and science, experienced as a lived contradiction in the case of the generation of Mori Ogai (1862-1922) and Natsume Sōseki (1867-1916), who came of age during its institution in the 1880s. It is only in this causal context that we can understand the fierce corporeality of Sōseki's rhetoric when, unable to reconcile the conception of "English literature" (*eigo ni iwayuru bungaku*) he encounters as a student in London with the conception of "literature in Chinese studies" (*kangaku ni iwayuru bungaku*) he absorbed as a youth, he decides that inquiring into the principles of literature by reading literature would be like "washing blood with blood," and chooses instead "science" as the mediation in which he can compare the fragments of his intellectual experience (Quoted in Komori 1995, 73-75). It does no violence to Sōseki's argument to understand science in the sense we use the term today. However literature is the site of a life-or-death struggle in the name of an older, more general sense, closer to our sense of the humanities today. If one accepts that the clash over literature and science is an example of an asymmetry characterizing intellectual discourse in the United States, the condensed form in which the university was instituted in Japan renders visible a symmetrically framed boundary between fundamental domains of human understanding.

There is finally the question of socialization. Though the establishment of a network of 53,000 elementary schools in 1872 made primary education accessible to every school-age child in the nation, these fed into 256 middle schools distributed regionally, and only seven higher schools and a single Imperial University at the top of a very steep pyramid (Saionji 1909, 162). With less than one-tenth of one percent of student cohorts going on to Higher School, by the 1880s virtually all future academic elite found themselves together for the common five year boarding school curriculum, with academic specialization not occurring until the Imperial University. The work of Donald Roden makes clear the crucial role this common experience in the Higher Schools played in instilling a ruling ethos in the future elite, detaching them like the English boarding schools from ties to family, local community and social class, and subjecting them to an intense period of socialization. Though the dorms retained for a time the roughneck, anti-intellectual character of the Kaisei Gakkō, the first article of the Imperial University Order of 1886 clearly charges: "the goal of universities should be the teaching of the arts and sciences essential to the nation." Higher schooling increasingly was tasked, then, prior to specialization in the university, with providing a common intellectual ground on the model of the French *écoles*, including a level of math and science literacy outside the scope of general education requirements in American universities. This unabashedly elite tuition meant that the research scientists and humanities scholars eventually produced by the Imperial University shared the intellectual and social background to communicate in spite of differences in their subsequent training path.[9]

Though the Higher Schools as a ground for socializing the elite were dismantled in the educational reforms which followed the Second World War, this common intellectual ground before specialization in the university survived in trace in a quirk of the examination system, whereby a serious run at admission to one of the elite schools that still produce the vast majority of professional scholars and critics in Japan presupposed a similar level of math and science literacy. Specifically, this means that for most of the twentieth century, scholars in the humanities in Japan have tended to be familiar with calculus. This has always been somewhat complicated by the difference between elite public universities, which required all applicants to test in math and science, and elite private universities, which allowed focused humanities students to substitute another humanities topic. However this situation itself appears to have changed in the last five years in Japan with a substantial revision of entrance exam guidelines and reduction of math requirements by the Ministry of Education. Asada Akira, who as author of *Structure and Power* (Kōzō to chikara, 1983) and leader of the New Academic movement in the 1980s, standing beside his piano in a spare apartment provided a model for the freedom of intellectual flight, speaks of an "appalling drop" in academic attainments among incoming students at his own Kyoto University. Specifically for the purposes of this argument, Asada, who makes his money as a cultural critic, pinpoints this in a drop in the feel for the way mathematical concepts and relations are mapped onto the physical world, which I have argued is the irreducible basis of science literacy.

> In terms of math and science, a person coming out of high school should be able, say, to solve differential equations, should understand that when something is thrown to the ground it describes

---

9. Murakami (1988), 72. This is clear from the sophistication and accuracy with which scientific and technical material was introduced in magazines geared toward middle-school readers at the time, such as *Shônen Sekai* (Boy's World) and *Yônen Zasshi* (Youth Times), available in the *Meiji Shinbun Zasshi Bunko* (Archive of Meiji Newspapers and Magazines) at the University of Tokyo.

a parabola, this is the absolute minimal common sense of modernity. In the new guidelines coming out, though, one can graduate from high school without knowing quadratic, much less differential equations. And despite that, in the interests of being 'up-to-date,' they've got a half-baked version of set theory in there. . . . I think it is rather more important that they be trained in the fundamentals of modern mathematics, from first and second order equations to differentials. (Asada and Fukuda 1999, 143)

The implied requirements for the elite course up to the point of these reforms provided for most of the twentieth century a common background for intellectuals in every field, and the grounds for an undiluted and mutually informed presentation of the concerns of science.

The Possibility of a Difference

The general points that come out of these considerations are, first, that the opposition between literature and science carries a symmetrical sense within the Japanese intellectual tradition in tension with the asymmetrical sense in which it is instituted in the university, second, that the participants in debate have a common background, and third, that a certain minimal degree of math competence is presupposed in the discussion. All this provides, in the unique context of early twentieth-century Japan, for both the intellectual tension to constitute a problem, and the common intellectual ground to address it in a symmetrical way. Though one could not from this *predict* a symmetrical engagement, it does suggest the possibility of a difference.

Writers in science studies confirm this intuition by turning to Japan to trouble putatively universal models of scientific practice. Sharon Traweek's *Beamtimes and Lifetimes: the World of High Energy Physicists* (1988) includes a report of substantial fieldwork in a Japanese particle physics lab that extended to after-hours discussions of the effect of the lack of plural construction in Japanese on the ability to conceive "particles" as individuated. Donna Haraway's classic *Primate Visions* includes a chapter on "multicultural primatology," where a host of tendencies in Japanese primatologists, including the provisioning of troops, employment of holistic rather than analytic reasoning process, extensive attention to gender as a component of social hierarchy, and cultivation of long-term emotional and cognitive connections between humans and animals, are found to escape point by point a western ideology of detached scientific objectivity, yet remain compatible with male domination of the academic field. And finally

Steve Fuller points precisely to the institution of faculties of Engineering and Agriculture in the Imperial University as indicating a healthy skepticism on the part of the Meiji leaders to the ideology of "pure science" being generated in the German and English universities that were otherwise their models. Because Japan in the late nineteenth century was itself under continual threat of colonial domination enabled by Western technological superiority, it could not, even for a minute, turn away from the instrumental aspect of science. Look at what the Westerners do, Ito Hirobumi advised his counselors as they picked and chose the characteristics for Japan's modern university, not at what they say in their myths (Fuller 1996; Haraway 1989; Traweek 1988). All these researchers go to Japan because of the sense that a fully articulated scientific sensibility is operating in an institutional and intellectual context uninvested in the idea of the universality of Western models. While respecting the methodological power of these first-rate practitioners of science studies, I can add to this body of work by accessing the intellectual groundwork in the original.

The point of reference within Japanese studies for situating science within the intellectual field is unquestionably James Bartholomew's *The Formation of Science in Japan*. This tightly focused study of the research tradition in Japan examines the formation between 1868 and 1920 of laboratories, research infrastructures, and recruitment of individuals, noting all major individual players, tracing the establishment of major institutions such as the Tokyo University science departments, Kitasato's Institute for Study of Infectious Diseases, the new Imperial Universities at Kyoto, Tohoku and Kyushu, and the formation and establishment of grants councils, giving close attention to the political and bureaucratic tussles at each major turning point.

The handling of factional struggles, personality conflicts, professor-student relations, etc. is densely detailed and highly judicious. While relying on social science terms like "values," "recruitment," "individualism" and "feudal legacies" to mediate Tokugawa institutions and early Japanese science, Bartholomew everywhere refuses the "culture" explanation, being concerned rather to see this institutional legacy with the Tokugawa period as enabling for Japanese science under great restraints (isolation, lack of funding). Consistent with his method, he brackets the epistemological question of science as a mode of knowledge production, presupposing an understanding of what modern science is, and arbitrarily but consistently defining "scientist" as a holder of the Ph.D. rank in a scientific field, with the presupposition that there was virtually no continuity at the level of ideas with the Tokugawa intellectual tradition:

> Despite changes . . . Tokugawa developments left their mark on modern science. Meiji (1868-1912) scientists did not draw on Tokugawa ideas, for they were almost entirely abandoned after 1868, replaced by ideas from the West. The Tokugawa contribution to modern science was not in the realm of the intellect, but in recruitment. (Bartholomew 1989, 4)

None of the Japanese historians of science I have quoted thus far (Maruyama 1961, Koyama 1991, Obayashi et al. 1994, Sasaki 1996), would accept that kind of statement. The two gaps in Bartholomew's study, then, which follow directly from his methodologically consistent restriction of "science" to work done by Ph.D.'s, are first in intellectual history, which he brackets, and second in the role of research in industry, which becomes an obvious problem in the Epilogue when he tries to extend the picture to contemporary Japan. These are not defects, though, but possibilities opened up for further research, and I hope that this study, in which questions of what science is, of what literature is, and of how different domains of knowledge are distinguished in the modern mode of production come front and center, will prove complementary to this indispensable work.

This leads to the following project: This book begins as it must, from close reading of the record of intellectuals in which I have glimpsed this struggle. The task from a literary critical perspective is to describe this phenomenon through detailed analysis of cases, and to explain by reference to independent variables in the system of culture of which it is a part. Because this system is a complex system, that is to say one composed of nearly infinite dimension and interrelationship between parts, and because this study is at some level a part of this system, such explanations are by nature partial, and the overall problem probably does not admit of a solution. The principle by which I have selected cases is as follows: First, all writers considered are major figures. Second, negotiating the boundary between literature and science is a central animating tension in their work. And third, at some point they kept the question of mathematical representation in the equation. Each of these writers is paired with a specific scientific issue, and analyzed in their most uncompromising scientific mode.

The study begins with a comparison of the pivotal figures of Ogai and Sōseki, and proceeds to an analysis of Sōseki's magnum theoretical opus the *Theory of Literature* (Bungakuron, 1907). Here

Sōseki presents a model of the flow of consciousness in reading drawn from the latest in contemporary cognitive psychology, which he works out in a masterpiece of impressionist literature *London Tower* (London tō, 1905). This model is mathematized in the first chapter with the famous {F+f} formula, however, the failure to include the equal sign blocked the ironic movement to a "quantification" of English literature. Yet Sōseki has larger quarry in mind, and the permutations implied in the pair of terms leads in a rigorous mathematical way to a fourth possibility, outside the domain of science. This is the realm he reserved for literature. The following chapter turns to Sōseki's most famous pupil, the physicist Terada Torahiko (1878-1935). Terada is best known in Japan for his voluminous, finely observed, and widely anthologized essays on the minutiae of daily life, and it is something of a truism that these eclipsed his scientific work. The reissue of his collected works by Iwanami Shoten in 1996 has had the effect of canonizing this aspect of Terada, to the exclusion of his scientific legacy. In keeping with the structure of this study, though, I will go straight to the science work, arguing that Terada through out his career repeatedly demonstrated a world-class physics intuition, finding himself on the forefront of research in X-ray crystallography in the 1910s, and decades ahead of his time in complex systems in the 1920s and 1930s. However, the failure in each case to push his insights to mathematical expression rightly haunts his legacy. Drawing on contemporary cognitive neuroscience research in creativity, I argue that his work in literature really did appear to work to the detriment of his science, calling up a different mode of creativity.

Chapters 4 and 5 move to the contemporary context to a pair of leading critics who use technical mathematical issues to write about early twentieth-century Japan. In the final chapter of his influential *Origins of Modern Japanese Literature*, Karatani Kōjin troubles the seeming naturalness of linear perspective by pointing to its status as a practical math heuristic put together through centuries of trial and error in the West. Hence for Karatani, the feeling of "depth" in painting and analogous feeling of psychological interiority in literature are alike transformations of sensibility through interaction with an essentially mathematical problem, "arbitrary as an aesthetic point." I then take up a highly enigmatic story entitled "Hell of Mirrors" (1926) by the mystery writer Edogawa Ranpo, and show that the seemingly arbitrary ending exploits this mathematical problem of linear perspective very precisely, and makes clear sense when the perspectival problem is taken seriously. Chapter 5 addresses the effort by the academic critic Maeda Ai to ground his readings of literature in the urban space in the mathematical field of topology. With Maeda we reach a peak of sorts, with a methodology promising a fully mathematical treatment of literary space. However, unfortunately, the use of topology does not stand up, and beneath the veneer of scientificity emerges a master practitioner of the ad hoc craft practice of literary criticism.

In each case, there is no fantastical juxtaposition of scientific and literary themes by analogy or participation in a zeitgeist. Rather, the scientific issues come out because they are explicitly discussed by the writers considered. Crucial to retaining the symmetry of the exchange is that I consider both scientists and writers as they negotiate the boundary between literature and science. I have reflected this in my choice of cases: Ogai and Terada are practicing physicists and physicians; Sōseki, and Edogawa are distinguished writers of fiction, and Karatani and Maeda are academic critics looking back on this period.

This book is conceived as the first of a three-part study, and in no way pretends to cover the subject. In particular, the discussion of Mori Ogai is deferred and will form a central part of a planned second volume. Second, science here is taken principally to mean the natural sciences, and aside from brief discussion in chapter 6, the interesting question of the social sciences is allowed to fall out according to the operational definition. Though I do not think any large problem inheres in the distinction between the natural and social sciences, historically in the Japanese case, Marxist debates around "literature and science" (*kagakuteki, bungakuteki*) in the 1920s and 1930s and Maruyama Masao's subsequent reconfiguration of the opposition in the 1950s to serve a polemical distinction between "Japanese" and "Western" attitudes to knowledge, cast the problem in terms of the social sciences, and require their own book-length treatment. The current volume is only a start, and while certain paths for future research are implied, I only hope at present to indicate the richness of the problem.

Any discussion of literature and science from a position in the North American academy must position itself in the barren terrain known as the Science Wars. Chapter 6 offers a detailed discussion in which I make the move, unusual in the humanities, of accepting the initial points raised by the science side, and argue that the impetus behind the controversy lies in overdetermined moves by the theoretically-informed humanities to erase the difference of scientific knowledge. There is a strong tendency to want to forget an episode like the science wars. However I would argue that, despite its nature as spectacle, the debate raised legitimate questions about the organization of knowledge in the modern university, that can only be ignored at the price of tacit recognition of the marginalization of the humanities proceeding apace in the United States. In the most substantial commentary on the science wars to date, Ian Hacking writes: "Far from wanting to sweep [the sticking points] under the carpet, I want to make it a central piece of furniture in the parlor of debate. . . . I do not want peace between constructionists and scientists, I want a better understanding of how they disagree, and why, perhaps, the twain shall never meet" (Hacking 1998, 31). Rather than react defensively, I would like to concede the point that workers in the humanities cannot talk about science without understanding science. But I have also noted that when the reaction against science studies reaches the antifoundationalist principles of the "theory-based" critique, that the science side falters, revealing a real issue at the heart of the skeptical attack, on which serious and conscientious people may be fated forever to disagree. Japanese thinkers less invested in the coherence of the project of knowledge found this rift long before.

A second unusual move is that I take science seriously. That is to say, I am not interested in challenging science in the domains in which it has authority, which is what I take to be the ethos, and quixotic task of much of science and technology studies. I am rather interested in *why* science has that authority and influence, and *not* the humanities. However, to inquire into the status of the authority and influence of science is not to grant it a claim to truth in any absolute sense, hence I am finally interested in the kinds of questions that are deferred, or excluded by science. These are additional questions that remain, even when the coherence and closure of science is granted with regard to the physical. They are *meta*-physical and phenomenal questions to which reason is driven by the nature of cognition and rationality itself. Let us take this adventure with these Japanese thinkers who, if I am reading this correctly, were attuned to *exactly* these questions.

The Science Wars can be said to represent a problem in the circulation of energies between the humanities and the sciences. The play by literary critics to incorporate science in the guise of literary theory short-circuits the relation, rendering it impossible to do work with the spectacular energy produced. Twentieth-century Japan is a site that can yield clues about how to reintroduce a ground in the form of mathematical representation, allowing the humanities to profit from the Science Wars, and not simply repeat it after a period of silence.

2   The Fourth Possibility in Sōseki's *Theory of Literature*

Object of little serious study in Japanese or English, Natsume Sōseki's *Theory of Literature* (Bungakuron, 1907),[1] with its mathematical formulae and extensive citation of English literature, and weighing in at over 500 pages, seems to be regarded as an eccentric indulgence by a writer who came later to be known through his creative works as Japan's greatest modern novelist. Or alternatively, the name is intoned with some reverence, but the theory left unengaged. Yet, opening its pages, one finds a scientifically grounded, fully articulated reader-response theory of literature, drawing on the latest in contemporary cognitive psychology, developed 60 years before the idea gained currency in Western literary theory. This chapter is concerned to elaborate Sōseki's model of reading, and to establish a tight link between the *Theory of Literature* and a contemporary creative work, *The Tower of London*, in order to place the *Theory* in continuity with rest of Sōseki's oeuvre. By establishing a link between Sōseki's magnum theoretical opus and his creative output, it is hoped to reposition this important work not as an optional appendage to Sōseki's literary oeuvre, but as an integral part of an original ten year plan to articulate the fields of science and literature in a general economy. It is in fact the mathematical formulae which provide the link.

---

1. Sōseki's work will generally be referred to by the translated title, *Theory of Literature*, however the Japanese title *Bungakuron* will also be used for economy and to distinguish it from shorter critical works with similar titles. The theory has not been translated, and occupies the entirety of the *Sōseki Zenshu* (SZ), volume 9.

The Situation of The Tower of London

A glance at the title, publication date, and first few pages of *The Tower of London* (London tō, 1905), might convince you that this early work, published shortly before the author's career as a popular novelist took off, is an example of the well-known genre of travel miscellany sent back in great numbers by the first generation of Japanese studying abroad in the universities of Europe and America. These short descriptive pieces performed the service of conveying to a Japanese newspaper or journal audience for whom the West was still largely a dream the details of travel, flora and fauna, the manners and customs of people encountered in traversing the globe, and the sights and sounds of the great capital cities at the height of the Western imperial project, and Sōseki is recorded to have published in the same year short pieces on entertainments, gardening, and bicycle riding in England, in such venues as the Imperial University Review and Japan Gardener's Journal (Kamei 2002, 243-244; Matsumura 1971, 547). *The Tower of London* was published in the relatively prestigious literary journal *Hototogisu*, and one might well imagine the interest and expectation the title reference to this icon of English history at the height of the Victorian era might hold for the contemporary Japanese audience. As we glean from the show of modesty in Mori Ogai's semi-autobiographical "The Dancing Girl" (Maihime, 1890), publication of these miscellany was an established avenue for the flow of up-to-the-minute information about the West, and a way for ambitious young men to make their first mark, even before the fruits of the their research became evident.

> It is now five years since the hopes I cherished for so long were fulfilled and I received orders to go to Europe. When I arrived here in the port of Saigon, I was struck by the strangeness of everything I saw and heard. I wonder how many thousands of words I wrote every day as I jotted down random thoughts in my travel diary. It was published in a newspaper at the time and highly praised, but now I shudder to think how any sensitive person must have reacted to my childish ideas and my presumptuous rhetoric. (Ogai 1994, 8)

Though identified in the West as novelists, Mori Ogai (1862-1922) and Natsume Sōseki (1867-1916) are positioned in Japan more broadly as central figures in the development of modern Japanese thought (Kōsaka 1958, 392-470; Karatani 1993, 11-44, 136-154). Both graduated from the elite Imperial University in the first decade of its consolidation, both spent an extended period of study abroad under government orders, and both cut their teeth as writers sending this kind of work back home. However, Sōseki's status as a government sponsored study-abroad student, or *ryūgakusei* was different in many respects from Ogai's. Unlike the prodigy Ogai, who was ordered to Berlin in 1884 while in his early 20s, Sōseki was already in his 30s and well along in a lackluster career when he was ordered to London in 1900.

Throughout the second and third decades of the Meiji period (1868-1912, i.e., roughly 1878-1897) the Japanese higher educational pyramid remained extremely narrow at the top levels, with a cut on the order of 10,000 to one between mandatory primary school and secondary levels, and a single national university at the top (Roden 1980). In 1894 only 341 students advanced to the university level nationwide, and the number of government sponsored students abroad remained around 20 during this period. Visible in legislation from 1897, though, is a systematic effort to expand the higher education structure, linked to the flow of reparations from the settlement in the Sino-Japanese war (Komori 1995, 59-60). A series of acts provided for designation of second, third and fourth imperial universities outside of Tokyo and the transfer of primary teaching responsibilities from visiting foreign lecturers to Japanese nationals. In 1897, the number of students promoted to the university level soared to 821, and similar measures doubled the number of *ryūgakusei* sent abroad by the Ministry of Education in 1900 to 39. It was this administrative expansion of the *ryūgakusei* system which caught up Natsume Sōseki at the age of 34.

Hence, when Ogai was ordered to Berlin in 1884 it was as part of an extremely select, top-class elite, sent to participate in cutting-edge research in medicine for a Japan still involved in a genuine struggle to secure its sovereignty against the Western powers. In fact, Ogai is credited with coining the Japanese term for research record (*gyōseki*) while studying in Germany (Sasaki 1996, 16). When Sōseki was ordered to London in 1900, however, it was as part of a major expansion of the *ryūgakusei* program and watering down of its prestige, with obvious connections to Japan's aspirations to parity with the imperial powers. Further, Ogai was sent to Germany to study his field of specialization, hygiene and medicine, while Sōseki was sent to England with orders to study not his chosen field of English literature but "methods and practices of English language instruction" (SZ 9: 5-8). Ogai's efforts abroad in the late 1880s were part of the good fight, the life and death struggle with the western imperial powers, while Sōseki's efforts in the early 1900s were part of the effort to become an imperial power. And though the significance is not immediately apparent, this positioned them differently in relation to the production of knowledge, and both would find their position in Japanese letters secured by what appears in retrospect as an almost unforgivable violation of the disciplinary organization of the university. Ogai was sent to study science and moved by a crisis of conscience to literature and philosophy, Sōseki was sent to study literature and moved by a crisis of conscience to study science.

Sōseki was clearly ambivalent about this situation, and the preface to *A Theory of Literature* records that he tried to refuse his orders halfheartedly, but was directed by his mentors to accept (SZ 9: 5). Uncomfortable from the start with his assigned role as a *ryūgakusei*, Sōseki resisted the constraints of this peripheral duty as travel guide as well. In the final chapter to *The Tower of London*, Sōseki expresses ironic concern for the disappointment of the reader.

> It might perhaps have been more convenient for the reader if I had given a more minute description of things in the Tower and how they were arranged. . . . But this piece of mine isn't meant to be a traveller's guide. In any case, it is some years ago that I visited the Tower; so I can't recreate the scenes very vividly. Consequently my account is somewhat overburdened by the repetition of subjective words. I fear this may give some unpleasant feeling to the reader; but it wasn't my original intention. I hope the reader will kindly take this situation of mine into account.[2] (Sōseki, 1992, 58-9; SZ v. 2: 29)

Sōseki calls up the genre of which this is a parody. The travel guide is characterized by factual information and concentration on minute description of buildings, places, peoples, historical points of interest, foods and other objects of tourism. This is what constitutes it as a genre, and Sōseki was aware his readers would likely approach the piece with these expectations. *The Tower of London*, however, departs from this in many ways that can only be labeled impressionistic.

---

2. Passages are taken with some modification from Peter Milward and Kii Nakano's fine translation in *The Tower of London* (1992). Page numbers from the English text will be followed by corresponding pages in the *Sōseki Zenshu* (SZ), v. 2.

The Tower of London begins with the narrator fumbling with a map, striding up to successive London street corners only to become completely lost again. At each corner he consults his map, asks pass-ersby or policeman for renewed directions and forges ahead only to become disoriented again at the next corner. After countless iterations, the Tower of London finally comes into view across the Thames, looming in the distance. The narrator feels pulled across the Tower Bridge by a kind of magnetic force, and once inside falls into a pattern of description very like the odyssey in the streets, setting out confi-dently with a description of the architecture, objects or historical facts concerning the Tower, only to lapse into subjective reveries that are close to hallucination. These passages typically summon up a tableau of historical persons imprisoned in the Tower, of condemned prisoners sitting in the boat bound for Traitor's Gate, of Sir Walter Raleigh at table in his cell writing his *History of the World*, of Lady Jane Grey blindfolded and led to the block by a priest. These reveries are pursued with Gothic intensity, accompanied by spilled blood and relentless speculation on problems of history, the irreducibility for the individual of annihilation by death and the possibility of knowing another mind. Each short chapter loses itself in this way for several pages, after which the narrator is literally brought back to his senses by a sudden sound, a breeze, the realization of another person in the room.

It is typical of Sōseki's irony that buried in the final apologies for the inadequacy of this kind of presentation is the crux of the problem. It is no idle request that we take this situation of Sōseki's into account, and inquire into the sense of subjectivity repeatedly produced by the text. We may begin by asking if it is possible to attach some sense to the intuition that the text is "impressionistic." The *Concise Oxford Dictionary of Literary Terms* gives: "Impressionism in the literary sense borrowed from French painting, a rather vague term applied to works or passages that concentrate on the description of transitory mental impressions as felt by an observer, rather than their external cause (Baldick 1990, 108). That is to say, it means one is concerned less with the object world, the outside world, real things, than with the flow of impressions they produce on the perceiving mind. Each chapter begins with the eye scanning an aspect in front of it, then makes an ambiguous transition to the narrator's reverie about the past. For example, chapter 2 begins:

When I first saw the Tower Bridge across the Thames, I was fascinated at the sight. I couldn't tell if it belonged to the present or the past. It was a quiet day in early winter. The sky was overcast, with a colour like that of liquid detergent in a pail. The Thames seemed to make its way downstream without a wave or a sound. . . . Below the Tower I caught sight of a sailing boat. . . .

As I gazed on the Tower, I found twentieth-century London gradually disappearing from my mind and giving place to a fantastic picture of the past. (26-27; SZ 2: 6-7)

Again in chapter 3, one begins with minute description of a visual aspect: "Crossing a moat, in which there was now no water, I saw twin towers ahead of me. They were round and made of stone, like an oil tank. They stood on either side of the path, like gigantic gateposts" (TL 29; SZ 2: 8). The narrator then comes to the Bell Tower:

I wondered how far the sound of the bell could reach. I looked up at the ivy-mantled tower and saw it had retracted all its sounds from past centuries and gathered them together in those of the present. (30; SZ 2: 9)

The bell signals the transition to the first extended reverie, as each successive detail of the Bell Tower suggests to the narrator something of the state of mind of the prisoners pulling up to Traitor's Gate. The reverie is brought to a close by the footsteps of a soldier on guard standing under the arch in front of him. With this the pattern is established of an ambiguous transition to fantasy, brought back and inserted into reality by the intrusion of a guard or another tourist.

Chapter 4 begins again with minute description of a visual aspect, "The walls of the Tower, I found, were built of irregularly cut stones, with a rough surface. In places ivy was growing up round them" (32; SZ 2: 10). The narrator moves inside, and in midst of describing the walls, bedposts and other items in a preserved chamber two boy's emerge from beside a bedside tapestry and a tableau ensues in which they leaf through an illustrated manuscript and discuss their impending death. The scene shifts to the mourners outside the gate, then to a pair of men dressed in black discussing the difficult points of their execution of the boys. As before, specific sensory impressions cause the collapse of the dream world.

Soon, another man in black seems to come out of the turret. . . . 'Till now, we have murdered many, but what we have done tonight surpasses all the rest in cruelty.' . . . 'When I strangled the boy there was a quivering of his lily lips.' 'The purple veins stood out on his clear forehead.' 'I can still hear his groan.'

Then they were swallowed up in the darkness, the clock on the Tower struck the hour, and I was awakened from my fantasy. The soldier, who before had been standing like a stone statue, was now marching up and down with his musket on one shoulder. (35-6; SZ 2: 13-4)

What emerges in the ambiguously marked movement between description of the external reality of the Tower and these elaborately observed fantasies is a consistent attention to the ebb and flow of the narrator's consciousness. Even in the transition to historiography proper, this is consistently maintained:

Entering the White Tower from the South side, I came to the famous armoury near the spiral stairs. There I found everything bright and shining, as they seem to take good care of each item. I was glad to see and ascertain for myself those objects of which I had only a vague idea from books of history or fiction. But I regret to say, my bliss of learning through such object lessons was of all too short duration, since almost all the knowledge I then acquired has since evaporated. (38; SZ 2: 15)

Sōseki's narrator has already made the ambiguous statement that, "The history of London Tower is the epitome of English history" (26; SZ 2: 6). However, what becomes clear from these passages is the instantaneous nature of the impressions, their ebb and flow, and the way Sōseki is constantly collapsing time in *The Tower of London* to the present of the observer. In the spirit of Berkeley's exploitation of the gap between mind and matter in a dualist epistemology, what is not present to the narrator's consciousness really does not exist (Berkeley 1952, especially 413-414). Within the framework of this consistently maintained impressionism, Sōseki goes on in turn to incorporate the different senses, sound, touch, binocular focus, which come flooding back in their concrete immediacy after each reverie:

A raven alighted near me with wings folded. He stared at me, with his bill ominously pointing at me. The blood shed here over hundreds of years seemed to have congealed and taken the form of this messenger bird. This is, no doubt, why the ravens can't depart from the site.

With such an impression in my mind, I looked up at the elm trees and felt the wind moving in them. (41-2; SZ 2-17)

Sōseki states here simply "one hundred years" (*hyakunen*). The translation "for hundreds of years" no doubt captures the historical sense more accurately, however as Karatani Kōjin points out in his extraordinary early essay, "Life as Viewed From the Interior" (1971, collected in Karatani 1991), we must understand the figure "one hundred years" to carry a precise significance in Sōseki's early works. In pieces like "Phantom Shield" (Maboroshi no tate, 1905) and "Ten Nights of Dreams" (Yume jūya, 1908), the appearance of the term "one hundred years" signifies neither "a long time ago," nor a lapse of 100 years in a chronological sense, but a leap outside life to the time before the birth or after the death of the individual, that is to say what is radically beyond the possibility of individual experience. Drawing step-by-step a coherent philosophical position from a seemingly incoherent mix of analepsis and prolepsis scattered over these works, Karatani concludes, "The term 'one hundred years,' consequently, indicates the leap to one's own death, and because this leap is beyond the possibility of experience for the individual (consciousness), it is simply expressed through the symbolic length of time, 'one hundred years' . . . [further], 'a hundred years ago' does not mean simply 'a long time in the past,' nor is it a concrete passage of time, it indicates rather the time before birth. . . . Death and birth are left virtually undistinguished in these 'dreams.' The more general philosophical point is that death and birth are indistinguishable when viewed from the 'interior' [of life]" (Karatani 1991, 16-21). In this sense the appearance of the enigmatic term "100 years" in *The Tower of London* clearly signals a departure from the domain of historical time measured in centuries, to the eternal present of consciousness.

> In tracing the thread of my thoughts thus far, I felt the cold atmosphere of the room seeping into the pores of my back, and I shuddered. I found it was moist in the room. As I felt the wall with my fingertips, I found it was dewy wet. My fingertips were all red. From the corner of the wall dewy drops were oozing out and dripping to the floor, making irregular red patterns. The blood shed in the sixteenth century seemed to reappear and wet the floor of the twentieth century. (47; SZ 2: 20)

> I felt spellbound in front of [Jane Grey's] inscription. Try as I may, I am unable to move an inch. I am again in a world of fantasy.
>   First, the eyesight grows dim and the things around me become invisible. In the darkness a light is lit. It gradually gets lighter,

and things take on clear-cut forms, as when one focuses on an object through binoculars. The scene becomes wider, and in the midst, I see a young woman sitting down. (52; SZ 2: 24)

That is to say, unlike a travel guide, and unlike a history, what is being recorded here is not the Tower, but impression, the ebb and flow of consciousness occasioned in the perceiver by the Tower. This is the stake of *The Tower of London*, and what pulls it out of the travel guide genre.

It would be easy to dismiss Sōseki's effort as romantic affectation, or to criticize it on grounds of historical accuracy, however, my argument is that this would be to miss the point of *The Tower of London*. I would argue that, far from being a "rather vague term," Sōseki's impressionism is an experimental embodiment of a precise sense developed in the *Theory of Literature* he was writing at the same time.

Sōseki's Theory of Literature

Considering the scope of the work, and its status as Sōseki's magnum theoretical opus, it is one of the puzzles of Sōseki studies that relatively little serious consideration has been given in either Japanese or English to the content of the *Theory of Literature* (Bungakuron, 1907). A number of Japanese scholars have traced the intellectual filiations of the work in detail, including (Shimada 1960, 1961) on the influence of Locke and James, and (Takahashi 1966) on the influence of Ribot. There has been a recent effort by Komori Yōichi to sound the topic in detail (Komori 1993; and 1995), and in English there is Matsui Sakuko's obscure but indispensable *Natsume Sōseki as a Critic of English Literature* (Matsui 1975). Matsui provides generous discussion of all the major critical works from Sōseki's time as professor at the Imperial University to his lectures for the Asahi news group, but tends to concentrate on Sōseki's characteristics as belletristic critic, rather than theoretician. Referring to the *Theory of Literature* as a "scientific and rather dry book," Matsui searches the work instead for points where "[Sōseki's] own likes and dislikes are noticeable, sometimes in spite of his own theories" (111-112). The first chapter of Karatani Kōjin's *Origins of Modern Japanese Literature* provides an accessible introduction to the work in English, concentrating on the strongly personal preface as symptomatic of the clash of sensibilities Sōseki felt in the face of English literature (Karatani 1993, 11-17). My reading of Sōseki proceeds from Karatani's work, and it is based on this account that Fredric Jameson counted Sōseki, along with Lu

Xun, as the major twentieth-century intellectuals to which Western theory has a blind spot (see Preface to Karatani).

Outside of these studies, sprinkled in various journals over 30 years, what reference does appear in mainstream criticism typically does not get beyond discussion of the highly personal preface signed with the given name Natsume Kinnosuke (*jo*, SZ 9: 5-17), in which Sōseki outlines in rich detail his near breakdown in England, and other unhappy circumstances attending the conception of the project. Even the lengthy commentary appended to the *Sōseki Zenshū* edition concentrates on the preface and Sōseki's personal circumstances, and spares less than a page to touch on the content of the theory itself (Komiya 1966).

As Karatani stresses, though, the *Bungakuron* as we have received it was originally part of a much more ambitious ten year project to situate the convergence of historically distinct senses of literature in the context of modern civilization (Karatani 1993, 11-17). The *Bungakuron* seems to have been meant as a first step, designed to rescue Sōseki from the disciplinary strictures of "literature" as he encountered it in England.

> Over the course of my study abroad I gradually came to dislike literature. Whenever I would read Western poetry, etc., I felt nothing. Trying to pretend that I was enjoying this would be like a person pretending to have wings and trying to fly, or a person with no money walking around trying to look prosperous. About that time, Ikeda Kikunae[3] came in from Germany and stayed at my lodgings. Ikeda is a scientist, but when I talked to him a bit, I was surprised to find that he's quite an impressive philosopher. I remember him besting me in arguments on a number of occasions. It was to my great profit that I met him in London. Thanks to him I was able to quit the spectral literature and formed the resolve to pursue a more systematic and substantial line of research. (Komiya 1966, 523)

Letters to his family from the period show the work in genesis and give an eerie sense of a person of preternatural sensitivity before whom something large and frightening has appeared, and he glances around to find no one else has noticed. In a letter to his father-in-law

---

3. Ikeda would later become famous for the isolation and patenting of the flavor compound monosodium glutamate, the commercial exploitation of which by the Ajinomoto company brought him great wealth (Bartholomew 1989, 180-181).

dated March 15, 1902, Sōseki describes himself being seized by a
new project, "reading day and night, taking notes and little by little
formulating my thoughts and advancing the enterprise. If I put out
the usual sort of book it's going to look like table scraps from the
Europeans, so I'm working diligently to produce something I won't
have to be ashamed to show to people" (quoted in Komiya 1966,
527). The preface describes the process of assembling books and materials for the project and the famous decision to "put all my literature
books at the bottom of a wicker trunk," the notion of trying to discern
the fundamental nature of literature by reading literature being for
the Sōseki of this period akin to "attempting to wash off blood with
more blood" (SZ 9: 10-11). Sōseki reports achieving a rare level of
concentration during this period, though his friends were apparently
concerned for his sanity, and he was reprimanded by the Ministry of
Education for failing to file necessary interim reports. Though he
expected the *Bungakuron* to take two to three years, the original conception was of a step-by-step triangulation of a number of disciplines
to produce a "dissection of the various elements structuring civilization, and discourse on its nature." "[E]ven I am shocked at the
scope of the problem" (Komiya 1966, 526-528).

Rejecting the methodologies of literary criticism as a way to come
to a fundamental understanding of literature, Sōseki resolved to approach the matter using the methods of the sciences, specifically
psychology and sociology. He formed the resolve to apply himself to
the utterly original question of by what indispensable conditions
psychologically literature can be "born into this world, develop and
fade away," and the rather more overdetermined question of by
what indispensable conditions sociologically literature "exists, flour-
ishes, and declines" (SZ 9: 10-11).

Sōseki's reading program is recorded in detail in the painstakingly
compiled *Bungakuron Notes* (Muraoka 1975), and there one finds extensive use of Havelock Ellis's *The Contemporary Science Series* for
access to the natural sciences and psychology, and Knight's *University
Extension Manuals*, given to him by one of his tutors in London. In
addition to these well-regarded collections, one finds as well a variety
of current works in psychology, history, ethics and aesthetics. Though
Sōseki's cross-disciplinary efforts in psychology accord well with
our conception of the scientific discipline today, as Komori Yōichi
points out, what Sōseki refers to as "sociology" seems more along
the lines of Spencerian history, ethics and aesthetics that formed the
fashionable intellectual currents of the day (Komori 1993, 36-38, 1995,

94-97). This tends to catch Sōseki up, when he turns to the sociological aspect of his project, in the company of such questionable figures such as Lombroso, Nordau, the more complicated Ribot, in generalizations from the individual to the social, and in social Darwinist themes such as Nordau's degeneration, and late nineteenth-century anxiety about entropy and decline, closely related to anxieties about the decline of empire (White 1990, 104-5; Hayles 1990, 214-5). The theory itself, elaborated with extraordinary analytical power, had to sit uneasily with the social Darwinism that underwrote it.

Sōseki can be said to have exposed himself, in his resolution to acquire the tools of the intellectual center, to a certain internalization of the Western gaze, which surfaces in moments of painful self-hatred in writing from the period, such as the famous description of surprise on seeing his own face reflected in the glass of a shop-window, and the angry close to the Preface of the *Theory* (See SZ 9: 14-15; Eto 1970, 22, 97-209). Komori Yōichi pinpoints the unavoidable ambivalence of a non-Western intellectual who would make a serious intervention into the thought of the center: "Caught up in the embrace of a natural science that had lent the power of its universal logic and evidentiary analysis to presenting the global hegemony of the European white race as historical necessity, and with a discourse of 'social Darwinism' that would eject and exclude any existence that does not conform to that order, Sōseki set out to raise the edifice of his literary theory" (Komori 1995, 96).

Upon his return to Japan in 1903 Sōseki was appointed in dual capacity to positions as English instructor at the First Higher School, and lecturer in English literature at Tokyo Imperial University, and the results of his research in London were delivered as a series of regular classroom lectures at the latter between 1903 and 1905. Sōseki had been hired to replace the popular foreign teacher Lafcadio Hearn, and the contrast between Hearn's theatrical romanticism and informal classroom style, and Sōseki's rigorous focus on methodology and first principles caused a near revolt among the students, leading Sōseki to great frustration and doubt about the project (Eto 1970, 241-257). At the same time, the unexpected success of a lengthy serialized novel *I Am a Cat* (Wagahai wa neko de aru, 1905) created a public demand for Sōseki's work. The *Theory of Literature* as we know it was collated and published in book form in 1907 as an attempt to capitalize on this growing popularity. Lecture notes were compiled and edited over the course of 1906 by his student, Nakagawa Mantarō (Komiya 1966, 538ff), and Sōseki was given the completed package for approval. A harried Sōseki had intended to sign off on the manuscript with a cursory glance, however contrary to expectations he found himself caught up in the task and took a full four and one half months for revision, from November 1906 to March, 1907. During this period, he ended up refusing all manuscript requests, and when he went into negotiations with the Asahi newspaper group to become their house novelist in February, letters to Asahi officials allude continually to a "bit of business" he is clearing up. Sōseki delivered the manuscript on March 27, 1907, and signed with Asahi the same day, refusing overtures from the Imperial University for a highly prestigious full professorship. This is the course of events from conception to completion that delivered the *Theory of Literature* in the form we know it today.

The editors of the annotated edition of the *Theory of Literature* found in volume 9 of the *Sōseki Zenshū* state that, though incomplete, the *Theory* is the single one of Sōseki's works into which he poured the most effort. It represents the indispensable preparation for his subsequent work as a great creative writer, and indicates the foundation that links his work as a theorist and literary figure. As such, Kadono Kiroku makes the case for Sōseki studies that, "to come to an appreciation of Sōseki's literary output, and to deepen research on his oeuvre, this stands as the crucial major treatise" (SZ 9: 549). Hence the problem remains of why so little serious research on the theory itself exists. Sōseki's highly rational, cosmopolitan *Theory of Literature* seems to be a category mismatch for a humanities scholarship organized by discipline and national tradition, fitting neither conceptions of Sōseki as a peculiarly Japanese novelist, nor conceptions of literature as a belletristic concern, such that it is difficult to recognize from the positions provided by the academic division of knowledge. There is the problem of the argument's extensive citation of English literature in the original, calling up competencies outside the purview of the *kokubungakusha*, or scholars of Japanese literature within whose discipline Sōseki studies usually falls. For the editors, however, this state of affairs is eclipsed by a different sort of cross-disciplinarity, the "uncanny feeling" given by the gesture to mathematics introduced from the opening page. According to Kadono:

> What puts readers off is likely this (F+f) notation. . . . Even a reader who picks up the *Bungakuron* with some anticipation is bound to be repelled when [just after the Preface] they run into: 'We can posit that the form of literary content be expressed as

(F+f) . . .', and the few lines that follow. As soon as someone sees that, they're likely to want to throw the book out. (SZ 9: 550)

What Kadono jovially suggests is that, for workers in the humanities, there is a visceral sense that nothing fertile can come from a mathematical formula. Hence the reading stops right there. This may not be true on an individual basis, but it does capture the reaction of Sōseki's students, and has some explanatory power in accounting for the relative neglect of this work. The question then becomes, why it was necessary for Sōseki to discard literature in coming to an understanding of literature, and embark on an interdisciplinary project of this scope, and how seriously we are intended to take the formulas, charts and graphs, and claim to scientificity announced in the preface, and sustained throughout the entire 500 pages of the theory proper.

And to do that, one must enter the *Theory* itself.

When Sōseki first arrived for his study abroad in London, he conceived his task to be to learn what methodologies must be acquired, and with which fields one must be acquainted in order to do first-rate research in literature. However, faced with an academic literary establishment in England that could not recognize as valid his own broad, situated understanding of literature, absorbed at the level of the body through the recitation of classical Chinese rhetoric as a child, and posited its own romantic notions of fiction and poetry as universally valid, Sōseki faced the crisis of consciousness alluded to in the earlier quotation. At this point, stimulated like Kant by the example of the natural sciences, he backtracked to first principles, and while professing embarrassment in the Preface for "going all the way to London to pursue such a childish problem," he formed the resolve to come to a precise understanding of "just what, at the most fundamental level, this literature is" (*konponteki ni bungaku to wa ikanaru mono zo*, SZ 9: 10). In Komori Yōichi's concise formulation, Sōseki makes the risk-filled turn to science as the mediation in which to adjudicate the competing claims "circulating within the head of a single Japanese exchange student" (Komori 1995, 94).

The first task for a scientific approach is to delineate the object, and in this Sōseki makes a surprising move. In attempting to grasp what, "at the most fundamental level," literature might be, he locates his object neither in the formal properties of the work itself, as in the Russian Formalists, nor in the activities of the writer as individual creative genius, but in the experience of the reader. Sōseki is concerned not with the thing, but with the person looking at the thing, hence the *Theory of Literature* is in the most fundamental sense a reader response theory. [4] He wants to develop a more general definition of literature because what he is being told is literature in England doesn't accord with his experience of Chinese literature, and he begins by building a model that will allow for a minute, moment by moment analysis of the experience of literature. The first chapter incorporates contemporary experimental psychological accounts of consciousness from Morgan's *An Introduction to Comparative Psychology* (1896) and E. W. Scripture's *The New Psychology* (1897) as a wave-like movement where the top of the wave is what the mind is focused on at a given point. Falling off in front and behind are both anticipation of future points and attenuated past points. This was a judicious choice by Sōseki as it does not appear that the basic model he borrows has been superseded in any important points today. The metaphor in use now by neuroscientists is of the conscious mind as a partially colorized movie, where what one is attending to at the moment is the part in color, but the rest is in black and white (Calvin 1999). Sōseki then brings forward a formula, the famous (F+f), incorporating concepts, impressions and emotions, and combines it with the model of consciousness to generate a model of reading replete with charts and graphs, that distinguishes the experience of literature from other types of conscious experience. The remaining 450 pages of the work then, is a thorough elaboration of all aspects of the model of reading he has put in place, even to the extent of dealing with "quantity of literary content," with reference to English literature.

Reading through the *Theory of Literature* as it stands is so deflationary to the pretensions of literature that one has to wonder if subjecting his chosen subject to this kind of unrelenting dissection is not a performance of sustained irony. In his discussion of the turmoil caused by Sōseki's succession to the lectureship in English literature at the Imperial University, in which a number of senior students stopped coming to class, Eto Jun writes:

> Sōseki's cold, penetrating analysis must have seemed to [Hearn's] students like a scalpel stripping the skin off the English literature that was like a lover to them. At any rate it did some damage to their pride. . . .
> It would be difficult to deny that in Sōseki was concealed an impulse to destroy illusion that bordered on revenge. He felt

4. Correspondence with Norman Holland (e-mail msg dated 12 Feb 2000 15:05:17 -0500).

challenged by the legacy of Hearn, and the students who blindly followed him imagining themselves geniuses, but there was also the desire for a thoroughgoing counterattack against the Englishmen and the England that had caused him such pain over the last two years. (Eto 1970, 250)

That Sōseki was traumatized by the rejection of his lived relation to literature, absorbed as a child in the study of the classics of Chinese rhetoric undoubtedly stands as a motivation for study (Karatani 1993, 11-22), and there is no gainsaying Eto Jun's insight into the matter. However this in no way obviates the need to press on to an analysis of the theory itself, which needs to be judged in terms of its coherence, and significance in the context of modern thought.

Terry Eagleton characterizes the development of modern literary theory as involving a major shift from a late nineteenth-century "preoccupation with the author," to an "exclusive concern with the text" in the early twentieth century with the Russian Formalists and New Critics. Examination of the reader's role in literature is understood as "a fairly novel development" occurring only since the 1960s with the advent of reception theory in Germany and the United States. "The reader has always been the most underprivileged of this trio" (Eagleton 1983, 74). Eagleton of course, knows nothing of Sōseki's *Theory of Literature*. In trying to grasp, "at the most fundamental level," what literature might be, the most obvious move for Sōseki in the first decade of the twentieth century would have been to seek the essence of literature in terms of the expression of the writer as individual creative genius. This would have been entirely comprehensible to the cadre of professors he faced in England in 1900. To have sought the object in the work itself as a self-contained formal object would have been ahead of his time, and made him a contemporary of the Russian Formalists. The *Theory of Literature*, though, seeks its object in the response of the reader, a full 60 years before the idea occurred to Western literary theory.

Taking a closer look at the model of consciousness Sōseki is adapting to the act of reading, we find the following charts and diagrams with the following explanations. The page will be translated in full to give a feeling for the style of argumentation:

The moment by moment activity of consciousness takes the form of a waveform, and if represented by a graph would look like below (see Fig. 2). As you can see, the summit of the waveform,

that is to say the focal point, is the most clear portion of consciousness, and before and after this point one finds the so-called
peripheries of consciousness. However, what we call our conscious
experience typically takes the form of a continuous series of these
psychological waveforms.

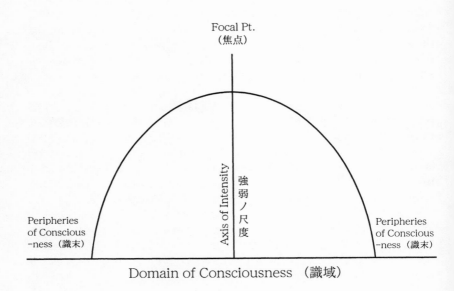

Fig. 2 Model of Consciousness in Sōseki's *Theory of Literature*

Following Morgan, we can represent this series as follows:

A	B	C	D	E	F	etc.
	a'	b'	c'	d'	e'	etc.
		a"	b"	c"	d"	etc.

That is to say, when conscious focal point [A] migrates to
position [B], [A] is attenuated to peripheral consciousness [a'],
and when [B] transfers further to [C], [a] and [b] take an additional
step toward the periphery of consciousness. . . .

    This is not just something we feel as part of our everyday
conscious experience, it has been demonstrated through exacting

scientific experimentation (I refer you to Scripture's *The New Psychology* for detailed explanation). (SZ 9: 30-31)

It would be easy to assume that Sōseki's reliance on late nineteenth-century sources vitiates the *Theory of Literature* beyond consideration today, however, the cross-disciplinary efforts in psychology are quite solid and surprisingly current. Sōseki was drawn to a materialist, cognitive psychology that sought to locate conscious processes in the experiences and stimulation of the body. The work was experimental, consistently critical of a dualist conception of mind and body, and the basic model Sōseki comes away with has been superseded in no important details in contemporary neuroscience. The soundness of Sōseki's intuition that the phenomenon of literature could be located in the second by second analysis of waveform of consciousness is echoed by neurophysiologist William Calvin, who writes, "[o]ne technological metaphor neuroscientists use for the enhancement of some images over others, so essential to our sense of self, is a black & white movie scene in which one actor becomes colorized, gradually standing out, until another actor develops color, and the first fades back to grayscale" (Calvin 1999), a conception entirely compatible with Sōseki's.

The model of consciousness is illustrated by the following example:

Let's say there's a person, and they're standing before St. Paul's Cathedral. Suppose that as they gaze upon that splendid architecture, their eyes move gradually from the pillars at the bottom section, to the balustrade at the upper portion, and finally reach the highest point at the tip of the cupola. While they are first gazing on the pillars, that portion of the structure is the only part perceived clearly and distinctly, and the rest only enters the field of vision indistinctly. However, in the instant the eyes move from the pillars to the balustrade, the perception of the pillars begins to attenuate, and simultaneously the perception of the balustrade gains in clarity and distinctness. The same phenomenon is observed in the movement from the balustrade to the cupola. When one recites a familiar poem, or listen to a familiar piece of music, it is the same. That is to say, when one separates off for observation a moment of consciousness from the continuity of a particular conscious state, one can see that the preceding psychological state begins to attenuate, and the portion to follow by contrast is gradually raised in distinctness through anticipation. (SZ 9: 30)

Let us return now to *The Tower of London* to strengthen the argument that the impressionism of this fictional piece is not a "rather vague" stylistic choice, but a precise embodiment of the model of consciousness developed for the *Theory of Literature,* and by implication for the coherence of Sōseki's project beyond the confines of the *Theory* itself.

One must first observe the similarity of St. Paul's Cathedral and the Tower of London first as monumental stone objects, and second as tourist destinations admired both for their architecture and relation to English history. Sōseki could have used a painting, a piece of music, the contents of his study, an apple, or a human face as the first object through which to illustrate the raw material of the model of consciousness he is developing, but instead selects an object that exceeds the perceiver by an order of magnitude in both size and longevity. The *Tower of London* begins in the same way with its narrator standing before a piece of monumental stone architecture. The first paragraph of *The Tower of London* opens as follows:

> Once while studying in England, I went and saw the Tower of London. Afterwards I sometimes thought I might go and see it again, but I gave up the idea. I was once asked by a friend to visit it with him, but I was unable to accept his offer. First impressions are too precious to be destroyed by the second. It is best to visit the Tower only once. (TL 23; SZ 2: 5)

The meaning of this recommendation, and the source of the narrator's puzzling confidence is a point much remarked on in commentary, and a riddle of almost unbearable obscurity to the translators of the English edition. In the Appendix to *The Tower of London,* Peter Milward gives an unguarded report of this consternation. "The first thing that catches the eye is, I would say, [Sōseki's] obstinate refusal to visit the Tower more than once. 'What an odd fellow!' is our instinctive reaction. 'Why won't he go a second time? Surely, what's worth doing once is worth doing twice, or three times. The more, the merrier! Chesterton even remarks that not until you've seen something a hundred times do you really see it for the first time.' But no! That is not Sōseki's idea. For him, 'first impressions are too precious to be destroyed by the second.'" (62-63) In fact, the translators of the English edition have done some work on this passage. The original passage reads as follows:

First impressions are too precious to be destroyed by the second. [And to sweep that away with a third is even worse.] It is best to visit the Tower only once.[5]

I have placed in brackets the sentence omitted by a translation otherwise characterized by a sensitive and conscientious adherence to the details of the original. Omitting dates, place names, description deemed repetitious is a liberty many translators of Japanese literature have felt justified in taking, however the omission stands out in this careful and stylistically superb translation. Like the good Englishman Sōseki parodies scrubbing a beautiful patina of moss off the rocks in his garden path, the translators dispense with the third iteration. 'Clear away that excess verbiage!' one might imagine the 'instinctive reaction' to be. But this point is crucial, as Sōseki is expressing here, in the first words of *The Tower of London* the diagram of successive moments of consciousness on page 31 of the *Bungakuron*.

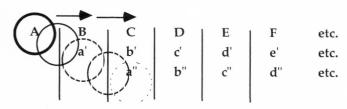

Fig. 3 Model of Consciousness in *Tower of London*

"First impressions are too precious to be destroyed by the second" corresponds to the attenuation of impression [A] to [a'] in the movement from focal point [A] to [B], and "to sweep that away with a third" demands to be read as the further attenuation to [a'']. The triple iteration, the successive movement of impressions, and the problem of attenuation clearly refer this to the diagrams of consciousness Sōseki is working on at the same time,[6] and the enigmatic insistence that one only visit the Tower once becomes, not a perverse

---

5. As I argue, the omission is not accidental, but signals a misrecognition of the stakes of Sōseki's account. The passage is as follows, with the omitted portion in brackets. "*Ichido de eta kioku wo nihenme ni buchikowasu no wa oshii, [mitabime ni nuguisaru no wa mottomo zannen da.] 'To' no kenbutsu wa ichido ni kagiru to omou.*"

6. The time in which *London tō* was written is not well known, but letters to friends indicate that Sōseki was working on the manuscript in early 1904.

restriction of the freedom of tourism, but a clear indication that he wants to keep this impression at the top, at the focal point.

Passage after passage, enigmatic on first reading, falls into place when the basis of the impressionism of *The Tower of London* in the model of consciousness in the *Theory*.

> It is only the scene of the Tower that stands out vividly in my mind's eye.
>
> When I am asked what I saw before that, I am quite at a loss. Nor can I say what I saw afterwards. In between those two spaces, where my memory is a blank, everything is bright. It is as if the darkness surrounding me was split with lightning under my very nose. Then the darkness returned. So the Tower becomes for me the focus of my worldly dream. (25; SZ 2: 6)

The narrator wants to retain a blank white space on either side of this vivid impression. Within the model of consciousness this indicates the preservation of the impression at the focal point as a singularity, so the imagery strains through the metaphor of the lightning flash or snapshot to eliminate the attenuated memory of past impressions and anticipation of future impressions. The Tower itself, the epitome of English history, is produced by the model as a singularity.

Passages quoted earlier show description inside the Tower maintaining constant attention to the flow of consciousness, to the narrator's mind as a desktop with limited space, such that the succession of impressions

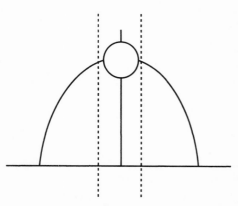

Fig. 4 The Singular Impression

while strolling through the Tower crowds out prior impressions, concepts and emotions, which trail away. Impressions and reveries flow and ebb, they "evaporate" and are replaced by new ones.

> As I gazed on the Tower, I found twentieth-century London gradually disappearing from my mind and giving place to a fantastic picture of the past. (27)

Hence I would argue that far from being a "rather vague" romantic performance, or a flawed presentation of English history, *The Tower of London* delivers a consistently maintained impressionism, describing not the Tower, nor the history of the Tower, but the transient mental states of the perceiving subject walking through it. It is an experimental work of precise design meant to embody in fictional form the model of consciousness Sōseki was working out simultaneously in his research on the *Theory of Literature*. I have spent considerable time establishing this point in order to link the *Theory* to the rest of Sōseki's oeuvre. In this way I hope to reopen the question of Sōseki's original ten year plan.

History, Memory, Impression

The central problem of *The Tower of London* then, indicated over and over by the collapsing of centuries into a single moment, and cued in the context of Sōseki's early work by the appearance of the cryptic signifier "one hundred years," becomes why Sōseki goes to such a historically saturated site to thematize a model of individual consciousness. Because of the need to bracket the socially and historically contingent, phenomenological accounts of conscious experience tend to take as their object of contemplation the ahistorical space of the philosopher's study, a desk, apple, cigarettes, pen and paper, the coolness of the paper in Merleau-Ponty's study, the back of one's own hand, and one could easily imagine Sōseki initially testing his model of conscious experience on something similarly close to hand as he wrote. Taking a public building as the object of consciousness brings up all sorts of higher order problems not easily handled by a rational introspection. By going immediately to a site almost hysterically permeated with social and historical associations, there are a number of points where, far from generating the figure of the contemplative intellectual, Sōseki sets up a position for his narrator very close to the one created for the Angel of History in Benjamin's mystical 1935 "Theses on the Philosophy of History," and the comparison is irresistible because Sōseki is ahead of Benjamin. And my argument is that Sōseki is ahead of the West.

The central point of the famous Thesis IX is the simultaneity of past and present for the angel, and the storm called progress. History appears to the angel as a single event, not a series of events, just as the Tower appears not as an object with a history but as a series of impressions in the present. The Thesis depicts the angel "staring, his mouth is open, his wings spread," and where we perceive a chain of events, "he sees one single catastrophe which keeps piling wreckage upon wreckage and hurls it in front of his feet." The angel would like to stay and do something, or at least inform and awaken someone, but a storm has got caught in his wings such that he can no longer close them. "This storm irresistibly propels him into the future . . . while the pile of debris before him grows skyward. This storm is what we call progress" (Benjamin 1968, 257-258). This may be compared to the opening of the mystical 2nd chapter by Sōseki:

> The history of the Tower of London is the epitome of English history. The curtain hiding the mystery of the past is torn down, and a beam of light comes through from beside the altar. Amid the gloom of the twentieth century it is the Tower which projects light. Time, which buries all in oblivion, flows back and comes to the present with the accidental wreckage of the past. The wreckage is what we call the Tower. (26; SZ 2: 6)

History is a wreckage. And both observers are horrified, but resigned with no possibility of intervening. The storm is what we call progress. The wreckage is what we call the Tower. And through a series of substitutions: The wreckage is what we call [the Tower]; which is the epitome of English History; which stands in for the modern West at the height of the imperial adventure. The angel is blown back by an irresistible force, while for Sōseki time, which buries all in oblivion, "flows back and comes to the present" in the figure of the Tower, an irresistible force which draws one in like a gigantic magnet. That is to say, what European intellectuals could not perceive about modernity until the wreckage of the 1930s, Sōseki grasped immediately, in reflecting on England at the endgame of its empire in 1900.

The meaning of the irresistible force called the Tower, figured as a gigantic magnet drawing the narrator across the bridge, is crucial to understanding the citation of the 3rd Canto of Dante's *Inferno* which closes the chapter: Abandon hope, all ye who enter here . . . [to the Tower] . . . which is the wreckage of history . . . which for Benjamin is "the storm we call progress." The allegory is quite clear for Sōseki, writing in 1904 during increasingly excited preparations for the war with Russia which would bring Japan to parity with the Western powers: "Abandon all hope, ye who *enter this history*.

Yet, enter this history Japan did, and Sōseki's seemingly offhand later lectures such as "The Civilization of Modern-Day Japan" and

"My Individualism" are concerned almost exclusively with the mechanism driving this history, and the possible positions one might take to avoid getting swept up in the wreckage (collected in Natsume 1992). Here is where we begin to perceive the necessity for Sōseki to elaborate his scientific and putatively universal model of consciousness at the historically saturated sites of empire, and how tightly this fits with the larger ten year project to produce "a dissection of the various elements structuring civilization, and discourse on its nature" (letter dated 15 March 1902). And we can also begin to perceive the connection of Sōseki's project to the front lines of contemporary European thought, for the question Sōseki is posing through his scientific scrutiny of literature, far from being an annexation of literature by science, is precisely that of what drops out when literature is sequestered as a trivial discipline in the modern university, and science taken as the only valid form of knowledge.

{F+f}

The question now comes up of the famous formula, by which Sōseki articulates his model of consciousness with the experience of reading. The *Theory of Literature* opens with the assertion that the content of the object of literature takes the form {F+f}, where F = impressions or concepts, and f = accompanying emotions.

> Let us represent the form of literary content by (F+f). 'F' here designates focal impressions, or concepts, while small 'f' designates accompanying emotions. Hence the above formula represents the combination of the dual aspects of impression and concept in the cognitive factor 'F', with the emotional factor small 'f'. (SZ 9: 27)

Komori makes a great deal of the collection of typically opposed categories of concept and impression under the single factor 'F', arguing that this introduces from the start a recursive instability in the formula with the factor 'F' wavering between two "utterly opposed poles" (Komori 1995, 96-99). However this is a misreading of what I take to be an essentially Berkeleyan frame for the theory. Berkeley's *A Treatise Concerning the Principles of Human Knowledge* opens:

> It is evident to any one who takes a survey of the *objects* of human knowledge, that they are either ideas actually imprinted on the senses; or else such as are perceived by attending to the passions and operations of the mind; or lastly, ideas formed by help of memory and imagination—either compounding, dividing,

or barely representing those originally perceived in the aforesaid ways. (Berkeley 1952, 413, italics in original)

An apple, stones, trees, a book and other sensible things usually understood to have a distinct external existence are defined in this way as "collections of ideas," which insofar as they are pleasing or disagreeable, "excite the passions of love, hatred, joy, grief, and so forth." This is precisely Sōseki's setup, and it can be seen that the surprise Komori professes at seeing Sōseki's union of concept and impression under the cognitive factor 'F' is rather premised on stereotypes of Western thought as ruled by the Cartesian dualism Berkeley is concerned to exploit. Whether or how Sōseki might have come in contact with Berkeley's writings is not known, however unlike Locke who is more typically read as an antecedent for Sōseki's thought but hedges on this score, Berkeley rigorously exploits the gap opened up between interior and exterior by the separation of mind and body in the Cartesian system. Berkeley takes the central problem of a dualist epistemology: If mind and matter are truly separate, how does one act on the other? That is to say, how can one know an inert external world, and answers in the most rigorous fashion, one can't.[7] The idealist problematic of an unknowable exterior comes out centrally in *The Tower of London*, *Ten Nights of Dreams* and other stories from the period, and the psychology Sōseki is borrowing is resolutely materialist, exploring a monistic system from the other side of the coin.

Clearly the system Sōseki elaborates is critical of a dualist split between mind and body, however I would like to read the stakes rather in the elaboration of the formula itself. Sōseki provides three possible permutations for the formula (F+f):

1) (F) but not (f)   i.e., pure concept or impression with no accompanying emotion
2) (F)   and   (f)   i.e., concept or impression accompanied by emotion
3) not (F) but (f)   i.e., pure emotion with no object (SZ 9: 27)

To which he assigns the three most convenient possible illustrations:

1) (F) but ~ (f)   the concept of a triangle, a pure concept

---

7. Berkeley goes on to save the common sense belief in the external world in the end by guaranteeing its continuity under the gaze of an all-seeing God, but this has rather the flavor of the pat moral affixed to the end of an anarchic fairy tale.

2) (F) and (f)   a flower, or the stars, concept/impression
                        w/associated emotion

3) ~(F) but (f)   dread, i.e., fear without an object (SZ 9: 27)

Literature, Sōseki says, and it must be kept in mind that this means "English literature," that is to say in its narrow sense as the modern university discipline, falls under case #2, i.e., concepts or impressions, accompanied by emotion, (F+f). He will spend the rest of the *Bungakuron* working out more ambiguous examples from the canon of English literature with the help of his students.

Sōseki then, has developed a formula incorporating concepts, impressions and emotions, and combined it with the model of consciousness to generate a model of reading that distinguishes the experience of literature from other types of conscious experience. The architectonic symmetry and fine attention to detail with which he elaborates the various quantitative and qualitative aspects of the experience of literature in relation to the formula (F+f) over the next 450 pages is breathtaking, and one must agree with Tsukamoto Toshiaki estimation that, "[i]n comparison with contemporary English books on rhetoric, The *Theory of Literature* is probably far above the average in its consistent logic, its fine argument, its rich sources and its concentration upon literature" (quoted in Matsui 1975, 111).

However returning to the very first page, one finds Sōseki has been strangely silent on a fourth logical possibility for combining F and f, namely the negation of both factors.

4) ~ (F) and ~ (f)   (neither (F) nor (f)); no concept or impression, no accompanying emotion)

Mathematical expression is a kind of best case of theoretical representation because it has an utterly clear syntax. Any combination of two elements has a total of four possible permutations, and it is strange that Sōseki neglected this fourth possibility for two reasons: First, the mathematical character of the exposition leads one irresistibly to posit it.[8] Second, Sōseki seems to have spent much in his creative oeuvre trying to approach exactly this fourth possibility. There are first the "others to life" (death, the time before birth) that Karatani sees obsessing Sōseki in his early stories and sketches, where terms like "one hundred years," or "the other shore" function not as chro-

---

8. Here the similarity to the manipulation of propositions in the truth-tables in Wittgenstein's *Tractatus* (1922) is clear.

nological or spatial quantities, but as signifiers for life as the gap
between two voids (Karatani 1991). Cases 1, 2, and 3, involving some
positive combination of cognitive and emotional states, obtain in the
span of life (*sei*), in the small gap of consciousness between two
voids, in the "hundred years." Case 4 obtains in death (*shi*) and the
time before birth (*tanjō izen*), in the realm beyond the possibility of
individual experience. The periodic interest of his otherwise thoroughly modern and alienated characters in the experience of *zazen* or
seated meditation, as in *Mon* and the 2nd Night in *Ten Nights of
Dreams*, again points to a grasping toward this utterly negative fourth
possibility.[9]

When you look at the *Bungakuron*, this fourth possibility fairly
leaps off the page, because of the mathematical setup. Why doesn't
Sōseki mention it within the pages of his theory?

Sōseki avoids having to deal with this fourth permutation within
the *Theory* because it is not verifiable by experiment. This is set up by
the qualifier that he is dealing with the experience of everyday life.
"The impressions and conceptions *we experience everyday* can be clas-
sified into three broad categories" (SZ 9: 27). Sōseki makes this qualification because he is developing a *science*, and the principle of all
scientific knowledge, the definition almost is that it must be tested
by experiment.[10] Sōseki's stated ambition was to produce a scientific
account of literature, and he was in continual contact with colleagues
in the sciences during the production of the *Theory*, and one of his
most famous pupils was a physicist. Hence it is not likely that he
took the label "scientific" lightly. The model is first of all rigorous,
and the elaboration with reference to English literature systematic
and comprehensive, an achievement remarkable to read even today.
As far as the stage of testing by experiment, Sōseki seems to have
regarded the lecture hall as a quasi-laboratory, and he would continually propose various ways of accounting for, quantifying and analyzing the experience of literature to his students, and then ask
them to consult the fact of their own experience to see if it does, or
does not agree with the model. This involves subjection of the inquiry

---

9. *Yume juuya* (1908) can be found in SZ, v. 8. A translation of the 2nd Night's
Dream can be found at <http://www.clas.ufl.edu/users/jmurphy/Yume10file/
Yume10ya.html>

10. This definition comes from Feynman (1995), 2. Norman Holland points out
that Feynman follows here earlier operational definitions of science by the physicist
Percy Bridgman (e-mail communication, 31 March 2000).

to a methodological formalization which relies on a community of trained professionals to verify insights achieved in the course of solitary reflection.

In this habit of interrogating the facts of his consciousness and submitting the results to the auditors for verification, Sōseki is in accord with accepted psychological practice of the day, and we can get a look at how this object is defined in William James's *Principles of Psychology* (1890), also delivered as a series of lectures and in which the notion of a "stream of thought" was first sounded.

> In this room—this lecture room, say—there are a multitude of thoughts, yours and mine, some of which cohere mutually, and some not. . . . My thoughts belong with my other thoughts, and your thoughts with your other thoughts. Whether anywhere in the room there be a mere thought, which is nobody's thought, we have no means of ascertaining, for we have no experience of its like. The only states of consciousness that we naturally deal with are found in personal consciousnesses, concrete particular I's and you's. (James 1952, 147)

However, this method of studying the mind was not peculiar to James, but current generally in the investigation of *normal* states of consciousness. The psychological literature of the day is filled with references to abnormal states, hysteria, hypnosis, automatic writing, split and secondary personalities, nitrous-oxide frenzies, etc. There is a figure called the *patient* who is the object of this kind of experimental inquiry, as symbolized by the university lecture hall filled with students watching. When the question becomes one of normal psychological functioning, though, a curious shift occurs, and the scientist and his student auditors become the object of their own inquiry. The viability of the equation of "facts" with the observation of one's own consciousness is clear in James's citation of a contemporary to begin his discussion of the constant change that attends the flow of consciousness:

> I go straight to the facts, without saying I go to perception, or sensation, or thought, or any special mode at all. What I find when I look at my consciousness at all is that what I cannot divest myself of, or not have in consciousness, if I have any consciousness at all, is a sequence of different feelings. I may shut my eyes and keep perfectly still, and try not to contribute anything of my own will; but whether I think or do not think, whether I perceive external things or not, I always have a succes-

sion of different feelings. Anything else that I may have also, of a more special character, comes in as parts of this succession. Not to have the succession of different feelings is not to be conscious at all. . . . The chain of consciousness is a sequence of *differents*. (Hodgson, Shadworth, *The Philosophy of Reflection*, quoted in James 1952, 149).

Hence, Sōseki's methodology, of interrogating the facts of one's consciousness and submitting the results to a community of specialists for verification was scientific, an accepted part of experimental psychological practice. What is original about Sōseki is the location of the object of literature in this way. Reading is, in this anti-Romantic project, a "normal activity," open to verification by experiment. Hence, a tension is installed at the beginning, in that the definition of his project restricts his analysis to those permutations of his theory which can be tested by reference to the experience of reading, however, manipulation of the mathematical formula in the theoretical setup insistently draws one to consideration of a 4th permutation for (F + f). Because his project is scientific, and the fourth possibility leads one to questions that cannot be posed scientifically.

He would explore the 4th possibility in his literature.[11]

Conclusion

Considered in itself, Natsume Sōseki's *Bungakuron* is an extremely impressive achievement, a highly elaborated reader-response theory 60 years ahead of its time. However, it is only in its place as part of Sōseki's larger ten year project that it can be grasped in its true ambition, for Sōseki seemed to be aiming to elaborate through parallel prosecution in fiction and theory a domain of thought heterogeneous

---

11. Roy Bhaskar's critical realist intervention in science studies highlights the specifically metaphysical burden Sōseki will place on literature. In *The Possibility of Naturalism*, Bhaskar rejects the idea that philosophy and science have different domains. "What is the relation between science and philosophy? Do they compete with one another or speak of different worlds? Neither position is acceptable" (4). Hence, Bhaskar's 'transcendental realism' rejects the opposition. In a clear allusion to Kant's discussion of science in the Preface to the Second Edition of the *Critique*, he writes, "Some implications of the conception of philosophy advocated here should be registered straightaway. Firstly, according to it, there is no connection between (a) what lies beyond sense experience and (b) some special sphere of philosophy . . . the familiar conflation of (a) and (b) in a unitary concept of metaphysics must be assiduously avoided" (7). Therefore one has to take seriously Sōseki's choice of literature, and not philosophy, as the domain for exploring the fourth permutation.

to the instrumental knowledge to which he reduces the entire western technological, political and military adventure. This reduction occurs in a lecture entitled, "The Civilization of Modern-Day Japan" (1911), delivered ten years after he conceived in a fever in London the ambition to produce a "dissection of the various elements structuring civilization, and discourse on its nature." In this later lecture the progress of modern civilization is defined as an interpenetration of energy-saving and energy-dissipating modes that is both pointless and always produces an excess. The model may be called thermodynamic in terms of its concern with transformations of energy within a system, but because there is no hint that the reservoir will be threatened with exhaustion by the irreversible process, this represents a clear break from the anxieties about entropy which conditioned the popular discourse in the late nineteenth century and consumed him during his time in London.

Elaborated with ironic simplicity in the 1911 lecture, Sōseki comes up, as the definition of civilization for which he makes his audience wait, with nothing less than an economy, necessarily productive of excess, relentlessly progressive, but with no brakes to the process or criteria to judge whether the relentless change is good or bad. In *Why Things Bite Back: Technology and the Revenge of Unintended Consequences*, amid the current climate of science boosterism, Edward Tenner warns that the long history of efforts to think the future missed something obvious about the notion of progress. "It is obvious that technology makes things better. By technology I mean humankind's modification of its biological and physical surroundings. It is also obvious that we are still unhappy with those surroundings, more discontented than when they were inferior" (Teller 1996, xi). Sōseki clearly discerned at the turn of the century this dialectic between excess and discontent as the essential, defining quality of modernity. The strength of a scientific model lies in its capacity to account for new circumstances which couldn't be foreseen by its author, and it is testimony to Sōseki's prescience that he could be transported to present day Japan or North America and would not have to be surprised by anything he saw. His deceptively simple thermodynamic system of an ungoverned process of technological innovation, and a population driven to ever more frenzied activity by its labor-saving contrivances is predictive of the ability of faxes, e-mail, the internet and other aspects of the current telecommunications revolution to divide our time into ever-decreasing units for more effective exploitation, and recent warnings from the New Economy about the alarming direction this inevitable progress is taking bring Sōseki's formulation from the beginning of the century to its endgame (Joy 2000). Sōseki could not have foreseen the content of the impending revolution in genetics, nanotechnology and robotics, but he grasped its mechanism precisely.

The possibility that behind Sōseki's ironic self-denial he grasped something essential about the process of modernity, and figured in his ten year project an escape across the boundary between literature and science as heterogeneous domains poses interesting questions of the homology with the other great early twentieth-century theorists of heterogeneity, Bataille and Bohr. The economy Sōseki produces in the context of a definition of civilization is uncannily close to Bataille's characterization of his collaboration with the physicist Georges Ambrosino as research in a general economy that would study "the movements of energy on the surface of the globe" (Bataille, 191). In relating Bataille to Neils Bohr's complementarity, another point in the early twentieth century in which the interaction between the scientific inside and outside produced an irreducible heterogeneity, Arkady Plotnitsky writes:

> According to Bataille, the general economy is a 'science'—a theoretical framework and a textual practice—by means of which one can relate the production, material or intellectual, of excesses that cannot be utilized. (Plotnitsky 1994, 19)

This chapter has been concerned to displace the properly scientific *Theory of Literature* from its current position as an optional appendage to his literary oeuvre to something central to a larger project of which Sōseki never lost sight. The question that needs to be carefully posed, in the context of careful scrutiny of these early twentieth-century explorations of the frontiers of knowledge, is whether Sōseki, in his negotiation of the boundary between literature and science and sounding of the problem of civilization, was aiming at a restricted, thermodynamic system based on notions of entropy, that is to say a restricted economy in Plotnitsky's sense, or whether he aspired, in advance of Bataille, to bring to clarity the heterogeneous systems of a general economy.

3 Do the Math: The Physics of Terada Torahiko and a Non-Reductive Creativity

And thus it is rare that mathematicians are intuitive and that men of intuition are mathematicians, because mathematicians wish to treat matters of intuition mathematically and make themselves ridiculous. . . . Intuitive minds, on the contrary, being thus accustomed to judge at a single glance, are so astonished when they are presented with propositions of which they understand nothing . . . that they are repelled and disheartened.

—Pascal, *Pensées I* (1670)

There is an amusing scene in *I Am a Cat* (*Wagahai wa neko de aru*, 1905), Natsume Sōseki's famous parody of the Tokyo intellectual scene, where the young physicist Mizushima Kangetsu comes over to the house of his mentor Kushami to try out a lecture he is to deliver to the Institute of Physical and Chemical Research that night on "The Mechanics of Hanging." Present are Kushami, like Sōseki a professor of literature at the imperial university, his friend Meitei, a dilettantish intellectual and man of leisure, and the housecat who serves as narrator for the events. Kangetsu begins by a tracing of classical references to hanging (Hebrew and Persian literature, Piers Plowman, Beowulf, etc.) and review of existing literature on the physics of executions, and proceeds to a full-blown mechanical analysis of the scene in *The Odyssey* where Telemachus hangs 12 of Penelope's maidservants for giving aid and comfort to the suitors. Meitei and

Kushami listen to Kangetsu's earnest presentation, make various bluffs about their competence in Greek, and are alternately childishly amused, titillated and bored. Then Kangetsu gets to the math:

> "Now, according to the theory of equilibrium pertaining to a polygon, which I believe you know, we have arrived at 12 equations:
>
> $$T_1 \cos\alpha_1 = T_2 \cos\alpha_2 \ldots . (1)$$
> $$T_2 \cos\alpha_2 = T_3 \cos\alpha_3 \ldots . (2) \ldots ."$$
>
> "That's about enough of equations," snorted my master.
>
> "But the equations are the whole backbone of my speech," pleaded Kangetsu.
>
> "Then how about giving the backbone, as you say, later on?" suggested Meitei, somewhat obligingly.
>
> "If we leave that out, my research on dynamics won't make any sense."
>
> "You don't have to be so modest. Just omit that part and get on with your speech," insisted my master without any concern."
>
> " . . . ."　　　　　　　　　　　　　　　　(Sōseki 1961, 83-88)

Beyond the general superficiality of academic repartee, Sōseki is of course parodying the reception in the humanities establishment of his own *Theory of Literature* (Bungakuron, 1907), whose inclusion of charts, graphs and mathematical formulae in the opening chapter, drawing on contemporary empirical psychology to model the flow of consciousness during the act of reading, consigned to obscurity an impressively articulated theory of reader response 60 years before similar work appeared in the West. The model for Kangetsu, the physicist frustrated at the effort to communicate his ideas across the boundary between literature and science, is widely regarded to be Terada Torahiko (1878-1935), pupil of Sōseki from his early days at the provincial Kumamoto Higher School, and professor of physics at Tokyo Imperial University from 1914-1935. When asked by Terada about a future course of study on the occasion of his promotion to the Imperial University in 1903, Sōseki, then occupying the chair of English Literature, famously advised his promising pupil to study something "universal."

> You should choose a universal subject. English literature will be a thankless task: In Japan, or in England, you'll never be able to hold up your head . . . (Miyoshi 1974, 57)

Sōseki's specific advice to his protege? "Study physics." Terada would accept Sōseki's advice, and with his mentor ensconced as a major figure in the humanities, go on to a successful career as a physicist. However, like Sōseki, he remained split and conflicted between the two throughout his career, and would ironically suffer the converse fate of having the apparently "literary" character of his scientific work cause it to be overlooked by a generation of physicists concerned only with hard science, while his lasting reputation in Japanese letters was secured rather by his prolific essays on the minutiae of daily life. It is Terada's scientific work, though, which held stubbornly aloof from the mainstream of physics at the time, that has recently been the subject of reevaluation in light of the emerging prominence of what is called complexity theory (*fukuzatsukei*) (Arima 1995; Ogawa 2000). Terada's adherence to a certain type of problem, non-linear, non-continuously differentiable, that had no hope of being quantified or handled experimentally until the advent of computers has something visionary to it that we can specify by saying it resisted verification by experiment. This arguably put Terada's problem-consciousness in a category with literary critical speculation.

However, to posit a literary quality to Terada's scientific work is not to seek a similarity in the picture of the worlds presented, it is rather a question of similarity in the non-linear process of creativity. It is well-known that Terada divided his energies almost symmetrically between his scientific work and his prolific essay writing. This has been understood particularly by the physics community as a failure of concentration that worked to the detriment of his science. This view is probably not wrong, but it is the argument of this chapter that before such a judgement can be made specific, a deeper connection needs to be sought, a ground on which the two endeavors can be compared. In a highly interesting attempt to bring two disparate fields of intellectual activity together called "Serendipity in Poetry and Physics," Mario Valdés, professor of Comparative Literature at the University of Toronto and Etienne Guyon, experimental physicist and Director of the École Normale Supérieure, argue that attempts to apply thematic comparisons between the disciplines, or to counter "poetic insights" in science with scientific rigor in criticism are bound to produce only "superficial glosses on the intellectual undertaking that is the human capacity to imagine" (Valdés and Guyon 1998, 28). They locate their object rather in a certain kind of mental activity shared between the two disciplines, a "playful looking for understanding or curiosity in new configurations of meaning" which constitutes the non-linear movement of creative intuition. Creative intuition is a good focal point for comparing distinct intellectual activities, in that it "brings together the many concerns that the educated imagination 'plays with'" and gives a relatively well-defined point in time over which the two activities can be compared.

Structuring a discussion of Terada's problem-consciousness in the sciences, his ultimately compromised effort to bridge the split with literature in a single career, and the fluctuating nature of his posthumous reputation, in terms of the question of creativity hints at how a disciplinary division of knowledge might rationalize over the division between literature and science a cognitive distinction between types of creativity, and produce through specialization "thoroughbred" intellectuals, highly trained in one or the other but unable to cross productively between the two. In this sense, an argument can be made for placing Terada in the company of other early twentieth-century precursors of complexity theory such as Bachelier, Lewis Fry Richardson, and the Scottish biologist D'Arcy Wentworth Thompson. The desultory investigations into growth, form, pattern formation and emergence that these scientists pursued appeared somewhat slow and antiquated in the context of the developments at the extremes of cosmology and particle physics taking place in the metropolitan centers of research. However, the gradual appreciation of the problem of complexity at the root of their work reveals the leisure enforced by perennially unsuccessful careers rather to have afforded them the defocused field of attention needed to make different kinds of creative associations, associations that allow them today to be celebrated as precursors of a science invisible to the reductivist mainstream of their day.

Organizing Terada's Work

According to Oguma Eiji, one of the editors of the revised edition of the *Terada Torahiko Zenshū* (Complete Collected Works, 1996) recently published by Iwanami Shoten, the work collected in its 30 volumes is usefully divided by genre into three periods: The first consists of the short stories and fictional pieces he wrote between 1899 and 1909 during his student days as a member of Sōseki's circle; the second is a period of about ten years after taking a position in the physics department at Tokyo University when his work for non-specialist audiences stayed close to science issues; and the third is the long period following his collapse from illness in 1919 when he turned to the prolific production of short essays on miscellaneous subjects (*zuihitsu*). Terada is best known in Japan for the work of periods one and three, however, according to Oguma, it is the work from period two between 1909 and 1919, consisting of lectures, translations, and essays reflecting on the adaptation of science to everyday life that most repays reading today.[1] What strikes one as fresh there, in his reflections on his specific area of expertise, is a consistent concern with the limits of scientific description. Given this, it becomes possible to discern a thread in the seemingly random accumulation of themes and topics of the essays and fiction of periods one and three, a concern to mark from different perspectives what lies beyond the capacity of human understanding (Oguma 1998).

At 30 volumes, the *Zenshū* imposes itself as the complete record of his oeuvre, and this body of work produced over 35 years is certainly enough to have occupied a lifetime for any writer. However, in Terada's case, it turns out this was a sideline, and none of the work he produced as a research physicist at Tokyo University is contained in the ostensibly comprehensive *Zenshū*. One needs to go to a separate, 6 volume *Scientific Papers* (1936), written entirely in English, to find the output of his main vocation. This is an interesting omission, unthinkable for a writer in any tradition prior to the seventeenth century, entirely in accord with the modern division between the humanities and the sciences.

Recalling Edwin McLellan's point that historians and critics of modern Japanese literature have tended to define their subject in terms of fiction, poetry, drama and "essays in the lyric vein" (Kato 1979), the expected move in taking up Terada from a position in literary criticism would be to talk about the finely observed essays and the romantic fiction by which he is known and loved, and which are a treasure trove of minutiae about the daily life of urban intellectuals and of details about famous figures circulating in the Tokyo literary world. In fact, the essays have been themselves conveniently gathered in a separate *Zenzuihitsushū* (Complete Collected Essays, 1991, 6 vols.) to answer to these expectations and facilitate this kind of discussion. Oguma Eiji, a social scientist reflecting on the work of editing the *Zenshū*, reacts against this economy when he recommends attending rather to the scientific essays. However, though thematically about science, the essays gathered in the *Zenshū* are journalistic and popular accounts, and translations, and fall comfortably into the marginal literary category of the essay. It is rather the *Scientific Papers* themselves,

---

1. These are contained in vols. 5-6; 10; and 14-15 of the 1996 edition.

published in journals of physics and the physical sciences over the course of 30 years, heterogeneous even to his "complete collected works" to which a study about the relation of literature and science that takes the term interdisciplinarity seriously must be drawn.

What will we find when we open up the dusty volumes of the *Scientific Papers*? One might expect a collection of quaint and dated experiments, long surpassed as scientific propositions and of marginal interest even historically. And indeed, "Terada Physics" (*Terada but-surigaku*) as it is popularly presented often seems like an "enlighten-ment" style effort to give non-threatening, accessible explanations in terms of existing science. It would be a mistake, though, to judge Terada's science by the way it is presented in the popular sphere. In fact there is a highly dramatic story in his *Scientific Papers*, and what has struck physicists recently and caused his reevaluation is that the phenomena on which Terada fixed his attention, turbulent flow, the mixing properties of fluids, sudden-onset phenomena and pattern propagation in nature, share a property that makes them of interest today: they are classical phenomena that are still not susceptible to scientific explanation, a problem of a different order but no less difficult than the non-classical uncertainty on which subatomic phenomena would soon be found to waver.

The question of creativity as a critical point will cause these two considerations: the heterogeneity of Terada's oeuvre, and the irreduc-ibility of the phenomena to which he was drawn, to fall out in an interesting way. It might be thought from the title of this chapter that I intend to reproduce the cliche that reductive thinking is not creative. Rather the notion of a non-reductive creativity necessarily implies that there is also a reductive creativity. Scientific and literary thinking alike share the moment of creative intuition as an essential part of their process. However, attention to the difference between reductive and non-reductive intellectual enterprises in terms of cognitive studies of creativity will lead the inquiry not to the wholeness and integration of literature and science Terada's career seems to promise, but once again to an irreducible gap expressed in the failure of the *Collected Works* to incorporate the *Scientific Papers*.

Placing Physics in the Japanese Intellectual Milieu

Terada Torahiko was born in 1878, and like all of his generation was a product of the transformation of the Meiji state with no direct experience of the Restoration. The architects of the Meiji Restoration, members of Kato's generation of 1830, immediately recognized in

Western learning one of the critical factors for instituting a modern state capable of competing with the West, and instituted over the course of the first two decades of the Meiji period, along with universal schooling, conscription, railroads, post and telegraph and all the other technologies of the modern state, a university system along German, French and English models. Science in Europe was itself undergoing a transformation in the nineteenth century, though, from the heroic age of the great discoveries of the seventeenth and eighteenth centuries, to the professionalization of science in universities as centers of higher education and research, and the sense of the term "science" that is relevant as a context for Terada should not be presumed.

In what we may take as the standard narrative of science in Japan, Sasaki Chikara divides science at the most general level into four types: The first is science in a cultural anthropological sense as the systematicity visible in the efforts of every culture to make sense of the natural world around them. Second is Classical science, by which is meant the severing of this investigation from myth by the Greeks and organization into empirical, mathematical and geometric branches. The third is Modern science, which indicates the coupling of this rational enterprise to certain historical and social conditions relating to state, religion and capital, and the fourth is Non-western science, which refers not to the relativistic systems of cultural anthropology, but to fully elaborated alternative empirical or rational systems, such as Chinese medicine, or the Tokugawa *wasan* method of calculation, which is the only other system to achieve a symbolic algebra (Sasaki 1996, 24-38).

In the category of Modern science, however, Sasaki identifies two separate revolutions. In the first scientific revolution, a number of conditions fell into place prior to the seventeenth century, including the diffusion of classical Greek texts in Greek and Latin translation, reports from the great age of exploration in ships, the revision of a religious basis of understanding with individual thought and reason, and the emergence of the state and rational management and manufacture techniques in modern capitalism. These gave rise in succession to the epoch-making realization by Newton that the laws governing the heavenly bodies and motion on earth were the same, the rendering exact of the Copernican hypothesis by Kepler, the development of the calculus by Newton and Leibniz, and the generation of a mechanical model of the universe, and trigonometry, by Descartes.

In this way mathematically, empirically, and in terms of its governing hypothesis of a machine-like universe, science in the sev-

enteenth century came to be systematized within the modern
state, which we call the classical systematization of science. And
beyond this, the notion arose that modern scientific rationalism
needed to be diffused throughout other fields of thought and
society in general, a movement called the Enlightenment. (Sasaki,
33)

The second modern scientific revolution occurs in the nineteenth
century, and involves the division of science into disciplines, and its
systematization in the modern university. In the first stage of modern
science, science tends to be led by practical activity, with exploration,
factory experimentation and the unevenly coordinated efforts of individual experimenters pulling science along in an unscientific way. In
the nineteenth century, with the institution in France, Germany and
England of research as a primary role for universities, science begins
to lead technology and the current distinction between science and
the humanities in the university on the basis of experimental method
and the curation of archives is largely in place. That the term "scientist"
was coined in the 1834 by the Englishman William Hewell is no
accident, and indicates the new consciousness brought about by the
division of science into specialized disciplines and its systematization
in the modern university (Sasaki 1996, 32-34).

France's École Polytechnic is the earliest example, and this trend
is inherited and elaborated by Germany after the Napoleonic wars.
The emergence of an institutional logic in the subdivision of fields is
embodied in the German universities, whose restriction of one chair
per field per university meant that the only way to add personnel
was to create new fields. In this process of differentiation and specialization that occurred throughout the nineteenth century, science held
onto its general sense as investigation into the principles of things,
while the term literature became progressively marginalized as one
sub-field among many in the humanities.

The architects of Meiji science and education had already passed
their formative educational experiences when they toured Europe in
the 1860s and 1870s, and the context they brought to the encounter,
namely Tokugawa period medicine and Chinese studies, was rather
more like the first stage of European science in the seventeenth and
eighteenth centuries, gentleman scholars pursuing the study of nature
as part of a broader program of cultivation. However, the model
they encountered and instituted in the 1880s was modern science
under its second stage, disciplinization in the university. By the time
Terada enters Tokyo University in 1899 and chooses physics as his

field of study, the discussion of literature or science or law presumes this division of knowledge into professional disciplines at a university. Japanese physics was formed in the 1880s as part of this larger institution of the university system, and the significance of this event from the perspective of the established discipline is nicely captured in the *Encyclopedia of Physics*, a reference geared for the professional physicist. In the context of very uneven development of physics outside of Europe, the authors of the major entry on the "History of Physics" find that, "[t]he most unusual implantation occurred in Japan. The Meiji regime that came to power in 1868 formulated a national policy of adopting Western learning. Japanese science students were sent abroad for extended periods of study and, upon returning to Japan, taught their specialty in a Western language" (Pyenson 1981, 411). There is a famous quotation by Nagaoka Hantarō, one of the first generation of students studying abroad, in a letter to his friend Tanaka-date Aikitsu studying in Glasgow in 1888. The quotation is originally in English, and gives an idea of the tension and isolation Japanese physicists felt, coming from a country still under the burden of extra-territoriality and unequal treaties, as they worked in the great centers of research in Europe: "We must work actively with an open eye, keen sense, and ready understanding, indefatigably and not a moment stopping. . . . There is no reason why the whites shall be so supreme in everything, and as you say, I hope we shall be able to beat those pompous people in the course of 10 or 20 years." By the time of the Russo-Japanese war in 1905, the tension of these early students vis-a-vis the West had largely dissipated. The authors of the article conclude: "Nagaoka need not have been so alarmed. Japanese physics had grown to maturity by 1922 when Einstein spent ten weeks in Japan visiting with his colleague and translator Ishihara Jun," and Werner Heisenberg lists as the nationalities represented at that rare convocation of genius under Niels Bohr in Copenhagen in 1924, "English, American, Swedish, Norwegian, Dutch and Japanese" (Heisenberg 1949, 111).

After doing preparatory work at the Fourth Higher School in Kumamoto, Terada enters the Physics Department at Tokyo University headed by Honda Kōtarō in 1903, i.e., midstream in this development, second generation, taught by Japanese physicists. Terada did his first graduate work on forcing disturbances in the periodic functioning of a system of geysers at Atami, Japan, and received the Ph.D. for a dissertation on acoustics and turbulent flow in the shakuhachi, a Japanese wind instrument made of bamboo (Terada 1938, v. I).

Though the Japanese higher education system was put in place using European models, a number of particular points made doing research there different from doing research in the centers of European learning. The first peculiarity lay in the stress the Meiji architects laid on practical learning at the expense of the liberal arts and pure research ideologies at the center of European higher education, embodied in the original Charter for the Imperial University in 1886 which stipulated that the mission of the universities should be the teaching of "the arts and sciences essential to the nation" (Murakami 1988, 72). Despite the drive to compete on an international level evident in Nagaoka's letter from 1888, the expectation for young researchers studying abroad was to bring in the latest developments from Europe, and correspondence surviving from the period up to the Russo-Japanese War (1904-1905) is replete with examples of students having to ignore or refuse orders from the Ministry of Education in order to make the most of their opportunities in Europe. Kitasato Shibasaburō (1852-1931), the internationally renowned bacteriologist who in the 1890s isolated the bacilli for tetanus, anthrax, and dysentery, had to refuse an arbitrary order during his study abroad in the 1880s to leave his post at the Koch Institute for a different university. He advanced his commitment to pure research against bureaucratic indifference throughout his career, and ended up founding his own institute outside the university. Similarly, Natsume Sōseki was ordered to London in 1900 to study not English literature, his field of specialization, but the practical "methods of English-language teaching," and had to ignore the order to accomplish the research that would lead to his *Theory of Literature* (Komiya 1966, 526).

In understanding the position of physics in the Japanese research complex, the flip side of this national pragmatism is that practical and applied scientific fields held a much higher formal position in the Japanese university system than anywhere else in the world with the exception of American land-grant universities. The Imperial University accorded central status to the faculty of law, and incorporated the faculties of engineering (1886) and agriculture (1890) while these fields were still considered appropriate to vocational and technical schools in Europe, a far-sighted move, innovative in the world context (Murakami 1988, 72; Bartholomew 1989, 93; Fuller 1997, 121-134). This emphasis and state support for practical knowledge in the Japanese university meant that the various sub-disciplines of medicine had built up world-class programs and facilities by the time Terada was studying in 1903, with the faculties of Agriculture and Engineering not far behind. However the pure physical sciences were still far behind the research centers of Europe in funding, facilities, number of researchers and public interest. With pure research not well supported in the universities, and no strong research ties to industry, Terada's receipt of the Ph.D. degree in 1908 put him in a surprisingly rare position. Of the 1360 conferrals of the Ph.D. degree between 1888 and 1920, over 75% were in medicine and engineering (656 and 366, respectively), with only 54 in physics, and 22 in mathematics (Bartholomew 1989, 51). This trickle of Ph.D.s was spread between the original department at Tokyo and an aggressive new department under Nagaoka Hantarō at the newly established Tōhoku Imperial University.

But though small in number, the physics community was highly talented. When Terada obtained his doctorate, he was only the 30th produced in physics by Japanese universities, one-tenth the number in medicine, and less than five percent of the total. Yet of the 26 prizes awarded by the Imperial Academy of Science up to that point, seven had gone to physicists, and seven to researchers in medicine, and researchers in physics were, at that point, already appearing in bibliographies of European science journals (Bartholomew 1989, 146-147).

After taking the degree, Terada received orders from the Ministry of Education to study abroad, and one is made keenly aware of the excitement of the age and the distance between Japan and the centers of physics from the fact that he was in residence at Berlin and Göttingen between 1909 and 1911, attending Planck's lectures in general physics, and returning home by way of the United States (Terada 1939, v. 1: x). Indeed, Terada's career coincides precisely with the golden age of nuclear science. He witnessed as a student Planck's revolutionary work on black body radiation in 1900 which severed the formula for power emitted by a body when heated from classical theories of thermodynamics and electromagnetism by proposing a quantum, or minimum amount of energy which can exist for any one frequency, and Einstein's explanation of the photoelectric effect in 1905, while events from Bohr's proposal in 1913 of the quantum theory of the structure of the atom to the final elaboration of quantum mechanics in Schrödinger's wave functions and the Uncertainty Principle in the years from 1925-1928 occurred during his time as a practicing researcher (Coughlan and Dodd 1991, 14-23). Terada's relation to this cutting edge, and the mainstream of normal physics by which it was supported is of the greatest interest in evaluating his subsequent scientific output, and is the subject of the next section. What can be concluded from this sketch of the academic situation of physics in Meiji Japan, though, is that by 1911 as a recipient of the Ph.D., member of the research faculty of Tokyo University, and by acquaintance with the cutting edge of the field through recent experience abroad in Germany, Terada was placed at the top of a highly competitive, elite hierarchy, but in a place poorly funded and removed from the centers of research in Europe.

Interpreting The Scientific Papers

The first point to establish in beginning our approach to the problem of creativity is that Terada had a first-class physics mind. Perhaps the best indication of his ability as a physicist comes shortly after his experience abroad, in his work on the Laue phenomenon and X-ray crystallography in 1913. By 1910, the significance of Einstein's trio of papers from 1905 on Special Relativity, the photoelectric effect and Brownian motion were beginning to work their way through the international physics community, and many ambitious young physicists, like Kuwaki Ayao, Tamaki Kijurō (mentor of Yukawa Hideki), and Ishihara Jun (who would study under Einstein and introduce the new theories), responded to the ferment by devoting themselves to the theoretical problems of relativity (Bartholomew 1989, 179; Koyama 1991). Terada, though, just back in 1911 from two years abroad, chose the study of X-rays. Looking back on this at the end of the century when X-rays have worked themselves into the fabric of everyday life, the choice might seem prosaic in comparison with the reconfiguration of space, time and matter implicit in the Special Theory. However X-rays had only recently been discovered, and for Terada to be doing work on X-ray crystallography in 1912 places him without question on the international cutting edge of physics. The standard narrative of X-ray crystallography traces Laue's elucidation in 1912 of the long-standing question of the nature of X-rays as a form of very short wavelength electromagnetic radiation, through the work of the English father and son team of the Braggs at the Cavendish Laboratory, who developed the theory and experimental method by which the wave nature of X-rays was used to determine the structure of crystals at the atomic level. For this Laue was awarded the Nobel Prize in 1914, and the Braggs in 1915. Terada's work on the Laue phenomenon between 1912 and 1913, for which he won the aforementioned Imperial Academy of Science prize in 1917, is judged by contemporary physicists as in no way inferior to the work on that phenomenon by the Braggs in England (Arima 1995, ; Koyama 1996).

Terada's interest in X-rays can be traced back to remarks in his middle-school diary from 1895 about Roentgen's discovery of X-rays. The nature of X-rays was not well understood in the decade after its discovery and two hypotheses were advanced in parallel: the theory that they were a type of electromagnetic wave like light advanced by Einstein and others on the continent, and the theory that they were an ultra-fine particle traveling at very high speed, advanced by the elder Bragg in England. Terada's work as a graduate student, which included work on tidal disturbances, ripples produced in mercury, and a Ph.D. dissertation contrasting the continuous manipulation of tones in the shakuhachi through variation of the angle of incidence of the lips with the discrete manipulation of tones in European reed instruments, gave him a great sensitivity to questions of resonance and wave phenomena. When Laue's paper, which conclusively demonstrated the wave nature of X-rays, arrived in Japan in the summer of 1912, he reacted sharply and immediately, evidently making the same connection as the Braggs, and setting to work immediately with a fraction of the laboratory support.

Terada's pupil Nishikawa Seiji describes his efforts at the Tokyo University Physics department. "At the time, we had only a couple of small, out of date X-ray tubes used for classroom demonstrations, and a couple of cheap induction coils to power them. Terada tried for a week straight, but couldn't get anything to come out on the plates" (Ezawa 1997, 351-352). Terada borrowed a larger X-ray apparatus from the medical department, and was finally able to duplicate the results and get the Laue spots to appear. Using a fluorescent plate rather than a photographic plate, he was able to observe the phenomenon in motion as the crystal was rotated, and arrived at the correct conclusion that the Laue spots reflected a diffraction pattern as the X-rays passed through the arrangement of atoms in the crystal lattice. Having arrived independently, and through a different experimental demonstration, at the same theoretical interpretation as the Braggs, he reported the result in 1913 in a set of three papers, including two in the British journal *Nature* (*Scientific Papers V. III*, 15-28). This was a brilliant demonstration of judgement, intuition and experimental economy, and a matter of public record, three months after the Braggs published. However, more than the lag time, what seemed to cost Terada recognition as independently arriving at the same conclusion was the failure to essentially connect the dots that led to the mathematical expression of Braggs' Law, a failure to follow through after a brush with greatness that colored his reputation among the community of cognoscenti, and was viewed with the keenest regret by a subsequent generation of physicists hungry for Japan to achieve international recognition (Arima 1996). Frustrated by the month or more spotted European scientists at the most crucial times by the need to ship papers surface mail to European journals, and finding the rivalries for authorship distasteful, Terada turned the work on X-ray crystallography over to his students, and turned for good from the mainstream of physics. In an essay summarizing the state of the field of X-ray diffraction for a Japanese journal in November, 1914, Terada does not even mention his own work (Nakayama 1997; *TTZ* v. 15, 30-39).

Stepping Out of the Current

The subsequent story of twentieth-century physics, how a consistent focus on universal questions of the fundamental nature of the atom and the disposition of forces and matter in the universe led the most brilliant minds of the century to forge ahead into domains outside the possibility of human experience, with the most unsettling and far-reaching results, is well-known. After flirting with the leading edge of this effort in his work on X-rays in 1913, Terada abandoned the dramatic mainstream, and turned his energy and experimental acumen to a desultory succession of problems that, considering the dramatic strides European physics was making at the time appear eclectic, and by the end of his career mundane and even dilettantish by comparison. To give a taste of the kind of problems he began treating, in the 256 papers in his collected scientific works, one finds considerable work on geophysics, earthquakes, tides and meteorological phenomena, propagation of the firestorm after natural disasters, capillary ripples and Lissajous figures, sonic qualities of various instruments, materials, breaking seawaves, pseudo-periodicities, dispersion patterns of ink in a viscous liquid, the structure of wind, fracture patterns in glass, propagation of sparks, relation of fracture patterns to natural patterning (egg division, animal patterns), rockslides and other sudden onset phenomena, shape of the coastline in Tosa, and on the physics of the falling of the camellia flower, leading to the frequent refrain in his scientific papers, "As for the literature concerning the subject, I could find none as far as my survey went."[2]

---

2. In *Scientific Papers* (1938), 6 vols. See "On the Capillary Ripple on Mercury

In addition Terada spent a great deal of time in the 1920s championing an obscure theory of Continental Drift by Wegener, largely rejected at the time by the European geophysics community (*Front* 1996, 6-7).

We may discuss a sample of experiments to get an idea of Terada's developing problem-consciousness and methods of investigation. For example, observation of the devastation of the Great Kanto Earthquake in 1923 prompted Terada to begin systematic study of the principles involved in the propagation of damage after a natural disaster. Following the earthquake, Terada immediately assembled his assistants and began a detailed survey of the damage, determining that the majority of the difficulty in containing the damage was due to the unexpectedly complex progress of the wind-driven firestorm. Lacking any means of direct observation, he constructed a simple experiment involving the use of aluminum powder in various liquids under temperature gradients. Pouring a mixture of aluminum powder and alcohol over a slightly inclined glass plate produced a temperature gradient due to the evaporation of alcohol in the moving liquid, resulting in a turbulent agitation which left an orderly, striped pattern on the glass. Close examination revealed a pattern of vortices to be occurring in the trailing edge of the progress, moving in the opposite direction,

---

Produced by a Jet Tube" [*Proc. Tokyo Phys.-Math. Soc.*, 1904], v. 1: 1-7; "On the Noise of Breaking Sea-Waves and Some Optical Analogies" [*Proc. Tokyo Phys.-Math. Soc.*, 1915], v. 2: 43-51; "Apparent Periodicities of Accidental Phenomena" [*Proc. Tokyo Phys.-Math. Soc.*, 1916], v. 2: 84-87; "On the Sound of Aeroplane and the Structure of Wind" [*Proc. Phys.-Math. Soc.*, 1922], v. 2: 258-262; "Propagation of Combustion in a Gaseous Mixture" [*Proc. Imp. Acad.*, 1926], v. 3: 130-132; "Experiments on the Modes of Deformation of a Layer of Granular Mass Floating on a Liquid—Some Application to Geophysical Phenomena" [*Bull. Earthq. Res. Inst.*, 1928], v. 3: 362-374; "Experimental Studies on the Form and Structure of Sparks" Parts I-VI [*Scient. Pap. Inst. Phys. Chem. Res.*, 1928], v. 3: 281-298; 333-350; 383-408; 427-444; v. 4: 1-26; 85-104; "On Gustiness of Winds" [*Rep. Aeron. Res. Inst.*, 1928], v. 3: 449-475; "On the Curvature of Islands Arc" [*Bull. Earthq. Res. Inst.*, 1931], v. 4: 244-249; "On Cracks and Fissures—Their Physical Natures and Significances" [*Scient. Pap. Inst. Phys. Chem. Res.*, 1931], v. 4: 84-87; "Water-Jet Affected by Tobacco Smoke" [*Nature*, 1932], v. 5: 83-84; "On the Motion of a Peculiar Type of Body Falling through Air—Camellia Flower" [*Scient. Pap. Inst. Phys. Chem. Res.*, 1933], v. 5: 123-137; "On the Mechanism of Spontaneous Expulsion of Wisteria Seeds" [*Scient. Pap. Inst. Phys. Chem. Res.*, 1933], v. 5: 151-158; "On Physical Properties of Chinese Black Ink" [*Proc. Imp. Acad.*, 1934], v. 5: 177-180; "Experimental Studies of Colloid Nature of Chinese Black Ink" Parts I-II [*Scient. Pap. Inst. Phys. Chem. Res.*, 1934-5], v. 5: 181-192; 256-272; "Crack and Life" [*Abstracts, Bull. Inst. Phys. Chem. Res.*, 1934], v. 5: 221-222; "Physical Morphology of Colour Pattern of Some Domestic Animals" [*Scient. Pap. Inst. Phys. Chem. Res.*, 1935], v. 5: 295-307; "Experiments on Radial Cracks Produced by Percussion upon Glass Plate with Initial Thermal Strain" [*Sc. Rep. Tōhoku Univ., Honda Anniv. v.*, 1936], v. 5: 307-312.

producing the channels. Meteorological researchers in the 1960s were taken by surprise when weather satellite photographs subsequently revealed exactly this pattern in snow-producing storm clouds (Kōchi Kenritsu Bungakkan 1986). It is a deceptively simple test of a vastly complex and powerful natural phenomenon, but the talent to make the simplifications and the assumptions in getting to the essential movement of the phenomenon is surely one of the marks of a scientific mind.

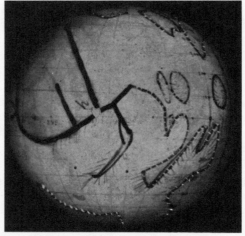

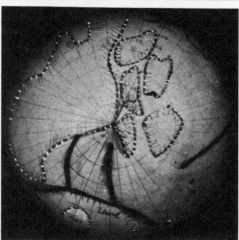

In "Physical Morphology of Colour Patterns in Some Domestic Animals" (1935), Terada goes through a complex set of steps mapping color patterns in animals (domestic cats, zebras, giraffes) with the pattern of division in cells, and crack patterns in concrete and other inanimate materials.[3] The intuition for the experiment came in observing his own black and white housecat. Terada traced the pattern of black and white markings on its fur onto a cloth in order to see how the pattern looks when pressed flat. He then cut the white pattern away from the

Fig. 5 Seam Lines from Color Patterns on Domestic Cats. (Scientific Papers V: 305)

---

3. See also the earlier *"Seimei to wareme," Scientific Papers*, v. 6: 286-287.

black ground, and bringing the tail and head portions together, sewed the pattern into a sort of pear-shaped pillow. The parts where the cloth had to be torn ended up looking like fracture patterns from inanimate material, which led to the following experiment.

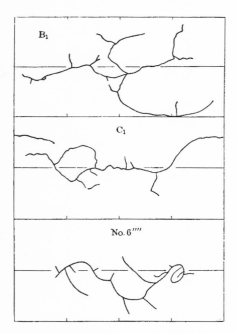

Fig. 6 Comparison of Seam Lines and Artificial Crack Lines

Terada mapped the tear patterns as faithfully as possible to a sphere, and records the consistent appearance of these patterns through the mapping process over numerous repetitions from a number of different animals (see Fig. 5). These sphere pattern were transferred using a mercator projection to a plane surface, and compared with the projection of similar crack patterns produced artificially in spheres of inanimate material, such as paraffin and plaster of paris (see Fig. 6). Terada found that these flat patterns in their propagation closely resembled the crack patterns in concrete, and hypothesized that, despite the infinite variation at the level of individual, they were falling out on a hexagonal grid that suggested the pattern of divisions in a dividing egg. This leads Terada to conclude that some *physical necessity* underlay these seemingly unrelated phenomena distributed over zoology, microbiology and materials science, and to hypothesize a mechanism for their transfer into living forms.

Several factors that run through Terada's later work are in evidence in these experiments. First is selection of phenomena on the scale of daily life and experience. Second is a reliance on modeling and simulation in a way that proves adequate in predicting large-scale complex systems. Third is a razor-sharp grasp of the interrelation of physical phenomena. Fourth is experimental economy. And fifth is a consistent interest in the production of patterns. His prolific work on geophysics, meteorology, turbulent flow, fracture and propagation patterns and sudden onset phenomena all share these characteristics.

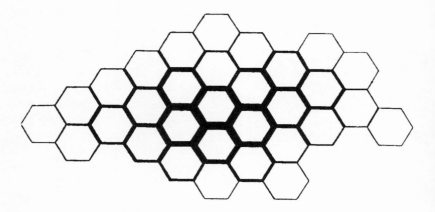

Fig. 7 Statistical Frequency of Seams and Intersections

The work Terada tended toward in his later years is qualitatively different from the work on X-ray crystallography, in that it lacks the orientation toward the fundamental structure of matter that seeks its problems through increasing levels of abstraction from the domain and scale of experience. Given that all physicists were not directly involved in the cutting edge of these theoretical problems, one is entitled to ask what kinds of experiments normal physicists were doing at the time. If one examines physics journals, one finds a large part of the energy of the field directed toward the isolation, wholesale accumulation and mapping of physical properties of matter, such as atomic spectra, "viscosity, elasticity, electrical and thermal conductivity, coefficients of expansion, indices of refraction" (Kleppner and Jackiw 2000, 893-894). This laborious, essentially empirical work, well-suited according to Kleppner to a Victorian work-ethic, was also well-suited to the large-scale mobilization of scientists and materials enabled by the big-ticket labs in Europe, and following in the wake of theoretical successes in probing deeper and deeper into the fundamental structures of matter, shared with them the tendency to reduce, quantify and abstract at the micro-level from the immediacy of experience. The success of this reductivist and descriptivist mainstream of science in the early twentieth century is beyond question. Terada's attention, however, was drawn rather to the action of complex systems at the aggregate, statistical level at which they are experienced and at which they again begin to show order.

Though Terada's later work appears like a falling away from the dramatic work on X-ray crystallography in the 1910s, a continuity is discernible with his earliest work before that brief interlude. An interest in the generation of statistically regular phenomena from turbulent, chaotic or complex action appears in his very earliest papers, though his treatment is at the macro-level of periodic regularity. There are articles on capillary ripples, and Lissajous figures produced by the interaction of two regular periodic functions on the surface of liquids, on acoustical notes, whistles, drum vibrations, the way small deviations in shape affect acoustic properties of the shakuhachi, on oceanic tides and the vibration of bars and wooden plates in various fluid media.

Though the thematic material of these experiments carries a quaint particularity, it would be a mistake to conclude that Terada's early thinking was not integrated into what Sōseki glossed as the universality of physics. Whether formulated in terms of a local hot springs resort, the tones of a shakuhachi or taiko drum, or a children's festival game, each of these shares the underlying problem of the generation of highly regular or periodic phenomena from the chaotic or turbulent excitation of fluids. Terada's first major study, "On the Geyser at Atami," published with Honda Kōtarō in 1906 in the *Publication of the Earthquake Investigation Committee* (*Scientific Papers*, v. 1: 49-68) is of some interest both for the way it foreshadows his use of things close to hand, and for the picture it provides of a young scientist modestly testing the waters in a representative set of problems. The experimental problem is to understand the forcing effect of wells drilled in the Atami area on the periodic functioning of a local geyser. The problem is interesting in that the mechanical problem itself is inaccessible because one cannot see underground and invasive examination would alter the object. Hence, Terada hypothesizes its operation, and constructs a mechanical model. If the model functions in the same way one has, as with neurophysiology of the brain, "explained" the phenomenon. Terada constructs out of vats, flasks and tubes a simulation that exactly reproduces the response of the Atami geyser to forcing in the form of well-drilling by local residents, thereby grasping the underground structure of the heating cavity of the geyser. His conclusion: "If the frequency of the eruption does not yet quite attain to its former value long after the stopping of the wells, we need not wonder at all, since some irreversible change in the subterranean mechanism might well have happened during the period of the disturbance" (66-67).

This tendency to choose problems close at hand, which appear mundane but are resistant to exact scientific description, is traceable throughout the *Scientific Papers*, and culminates in the set of papers on Chinese Black Ink from 1934-35, and on "[T]he Motion of a Peculiar Type of Body Falling Through Air—Camellia Flower" from 1933. The problem of Chinese black ink, or *sumi*, involves the elucidation of the mechanism behind the festival activity of *suminagashi*, similar to the spin-art of American carnivals, wherein several drops of black ink are dropped on the surface of a container of water, then transferred to heavy paper producing a beautiful design that can be carried home as a keepsake. Here Terada brings 30 years of physics experience and the resources of the Tokyo University Physics Department to bear on what is essentially a children's game. The latter investigation of the dynamics of the death throes of the camellia is clearly a follow-up and homage of sorts to his mentor Natsume Sōseki, whose novel *Sorekara* (1911) begins with the famous sequence wherein a camellia falling to the tatami mat insinuates itself into the dream of the protagonist. The blossom, "nearly as large as a baby's head," becomes an image of death whose oppressive presence drags throughout the subsequent novel (Sōseki 1978). The scene is brilliantly executed, weaving together precise description of the natural phenomenon and the literary world of dreams, and Terada's investigation of the science behind the drop is impeccable and not without its own irony, but when one considers the extraordinarily dramatic strides European physics was making at the time investigating and elaborating the new theories about space and time, and the fundamental structure of matter, these experiments look folksy, almost embarrassing.

And that is exactly how they appeared to the next generation of scientists in Japan. Terada's tendency to select the content of his work from everyday life around him carries what might be called the "taint of Japanese particularity." This grated on the sensibilities of young postwar physicists for two reasons: First, discussion of phenomena familiar from Japanese daily life, like his dissertation on the acoustics of the Shakuhachi, studies of Geysers at Atami, etc., appears hopelessly mundane compared to the drama of nuclear physics and cosmology unfolding in Europe, and second it appears as a withdrawal and retreat into Japanese particularity in the face of the universal claims of science. Terada's treatment of these problems is impeccable from the point of view of scientific method, reductive, quantified, tested by experiment. Yet who is possibly going to verify, contest, or even take an interest in claims about sugar-plums, or a children's game of ink patterns? Hence, for an ambitious generation of young physicists, galvanized by the Nobel Prize-winning work by Yukawa Hideki on the pi-meson in the 1930s and Tomonaga Sin-itiro on quantum electro-dynamics in the 1940s, and anxious to advance to the frontline of theoretical physics themselves, Terada's legacy worked against the tendency toward exact, universal knowledge physics required, while his skittish movement from one promising problem to another before they were fully elaborated, and the very success of his prolific sideline in essay-writing, where his interests went as afar afield as poetry and film, seemed evidence to them of an unpardonable squandering of physics talent. One may recall in this respect the urgent tone imparted to Japanese physics in the letter from Nagaoka Hantarō of 1888, and the aggressive character of the program at the rival Tōhoku University he was tapped to chair during Terada's career. Nakayama Shigeru writes of the "Yukawa effect," where Terada's physics, epitomized by the sugar crystal and suminagashi papers, was sneered at by Tokyo University students ("that stuff isn't physics"), who arrogantly refused the sweets Terada passed out in seminars (Nakayama 1997, 1). Particle physicist Arima Akito speaks for the succeeding generation of Japanese physics when he recalls:

I learned my statistical methods and experimental physics in college from one of Terada's pupils, Hirata Morizō. Eventually, however, I went into particle physics. At the time, Terada's work appeared to me and everyone else to be dilettantish. In particular, we couldn't stand the work on *suminagashi*. (Arima 1995, 26)

Reevaluation of Terada's Work in Light of Complexity Theory
In fact, Terada's posthumous reputation in Japan has rested almost exclusively on his status as an essayist, frequently anthologized, beloved of the general reader, a question of killing with faint praise for a scientist. But stirrings of a reevaluation of his work have emerged in both the scientific and general intellectual community in the last five years, prompted by two factors: First is the reissue of his complete works by Iwanami Shoten beginning in 1996, and second is the emergence over the last decade of a set of strategies for handling complex,

that is to say non-linear, dynamic, interdependent, or non-continuously differentiable, systems in the natural sciences grouped under the rubric of complexity theory.

The basic insight driving complexity theory may be said to be the observation, deferred for most of the history of science, that non-linear and non-continuously differentiable phenomena are the rule in nature, not the exception. In this sense complexity theory can be grouped with better defined fields like chaos theory and fractals as strategies for dealing with non-linear phenomena. Of course, the problems of complexity have always been there, in classical mechanics, thermodynamics, vibrations, etc., well-recognized, but unable to be addressed at the micro-level, because the capacity to specify initial conditions, and the tools to handle the extremely unwieldy calculations they produced simply did not exist. Hence while the ubiquity of non-linear phenomena was recognized, they could only be handled statistically, treated as special cases of exact or linear phenomena, or approximated as linear over short intervals. What has brought these theories into the foreground in the last 15 years is the advent of high-speed computing in the 1970s, which has for the first time made them susceptible to engagement.

Chaos and fractal theory, as opposed to complexity, are relatively well-defined mathematically. Chaos relates to the behavior of non-linear systems in a manageable number of dimensions.[4] If a linear process is one in which changes at some initial time produce proportional changes at some later time, in a non-linear system small changes in the initial conditions can lead to very large and disproportionate swings in the value of the system at later times. This leads to the canonical illustration in meteorology of the "Butterfly Effect," where the flapping of a butterfly's wings in South America produces a tiny disturbance in air currents that results in a typhoon in East Asia six months later. What makes linear processes so attractive in modeling natural processes is that they can be analyzed. A linear equation of whatever complexity and size can be broken up into a series of smaller problems, each of which can be solved separately and then the individual results combined into an overall result. A non-linear system, on the other hand, cannot be analyzed into a set of smaller problems, and must be solved as a whole. This gives linear systems a tractability, and the tendency of non-linear systems to behave in an approximately

---

4. The following discussion draws primarily from Venkataramani (1997) and Holden (1986).

linear fashion over short ranges has allowed the question of non-linear systems to be deferred for most of the history of science, despite their predominance in fact. Fractal geometry shares with chaos a reluctance to accept the move that would simplify the infinitely complex phenomena of nature in order to allow a tractable expression in mathematics. Fractal geometry takes the shapes and forms of nature, such as dusts, biological forms, coastlines, and rather than reducing them to some approximation of ideal forms of geometry in one, two or three dimensions, locates complex scaling on all levels that gives them continuous dimensionality. These are expressed in a mathematics that is exact, but neither linear nor continuously differentiable. The irregular, complicated solutions produced by the strategies of Chaos and Fractal geometry have elucidated the mechanisms underlying hitherto unmanageable physical processes, and have become important in fields as diverse as meteorology, cellular metabolism and population biology.

Complexity theory is not yet well-defined as a science, but in contrast to chaos deals with systems with very large dimensionality, that is to say, many interacting parts. Classical statistical problems such as turbulent flow or phase transitions in fluids are sometimes taken as the benchmark problem, sometimes complex adaptive systems that respond to forcing factors by finding another equilibrium, like the brain or ecological systems, or human society, and sometimes the problem of pattern formation and structure formation in the propagation of complex systems. In all cases, the complexity, or high level of interacting parts and interrelated factors means that it is difficult to specify initial conditions or describe the system in a reductive way. The congeniality of this description to phenomena like "culture" or "society" is immediately apparent to the worker in the humanities. The interrelatedness of the emergence of computers and complexity theory is indicated by information theory, in that there is a tendency to define the complexity of a system in terms of the amount of code needed to produce a model that successfully approximates its inputs and outputs, a project that must give the cultural critic pause and recalls the information theory definition of randomness.[5] In all cases, chaos, fractals and complexity theory, the practical emergence of the field had to wait for the advent of high-speed computers.

---

5. A random number is incompressible, that is to say it cannot be generated by an algorithm that is shorter than the number.

An argument has begun to emerge in the reevaluation of Terada's scientific achievement occasioned by the reissue of his collected works, that the problem with Terada's physics from the perspective of complexity theory is not that he chose obscure Japanese topics that were too particular to be recognized by European science. The problem is that he was 50 years ahead of his time, and the field of physics had to wait for the advent of the computer to understand the purport of his questioning. In this argument, advanced influentially by Arima Akito, a particle physicist, former president of Tokyo University and Director of the Institute for Physical and Chemical Research, the interest in the reevaluation of Terada's scientific work comes from the perception that the type of problems he stubbornly attended to throughout his career, despite their seeming inconsequentiality from the point of view of the mainstream of physics in the 1920s and 1930s, turned out to be remarkably prescient of the types of problems that have showed themselves amenable to the new sciences of Complexity (Oguma 1998; Arima 1995).

The properties in Terada's work and problem-consciousness that distinguished him from the mainstream of physics in the 1920s and 1930s, and anticipate complexity theory for contemporary physicists seem to be the following: First, selection of objects on the scale of human experience (versus the extremely large, extremely small, extremely fast or heavy); selection of problems that are classical, but insoluble, i.e., a statistical uncertainty, not a quantum uncertainty; and finally, an uncanny tendency to stumble on the classical locus of complexity problems, in phenomena of turbulence, patterning and sudden onset that are still highly resistant to scientific description.

The interest in the generation of statistically regular phenomena from turbulent, chaotic and complex action is present in his earliest papers, though his treatment is at the macro-level of periodic regularity. In his later years he began to link this interest in the turbulent excitation of fluids to the question of phase change and critical points, and further with the production of patterns in highly dynamic propagating systems, spark discharge, fractures, patterns in crystal and organic growth. All these problems are basic breeding ground for recognizing the problems of complexity.

More specific citations of Terada as a progenitor of the new studies of complexity come from a working group of natural scientists and engineers assembled in 1985 from various fields called the Society for the Science of Form. The science of form is a translation of the Japanese term *katachi no butsurigaku* (physics of form) or *katachi no kagaku* (science of form), and is translated in English variously as "morphological science," "stereology," and "science of pattern-formation." All these variations share a concern with the geometrical nature of structure, the mechanisms of pattern formation, the quantification and measurement of forms, and the intersection of this science with art and design (Takaki, Arai, and Utsumi 2000, 1). While finding classical antecedents in Kepler and 16th century drafting, more broadly implicit in the work of the Society, which includes participation by physicists, mathematicians, statisticians, biologists, biophysicists, geographers, architects, anthropologists and city-planners from Japan, North America and Europe, is the notion that the Japanese term *katachi* has a polysemy that the English term "form" lacks, and that this favorably conditions Japanese scientists in their approach to the problems of complexity (Takaki 1997; Ogawa 2000). Reference to Terada Torahiko's work is pervasive in the introduction to the proceedings of the 1995 symposium on "Katachi U Symmetry" (read Katachi 'union' Symmetry), and his work on fracture patterns, propagation of lightning, sparks, X-ray diffraction and stripe patterns in animals is placed by the editors in the same category as D'Arcy Thompson's pioneering work on the physical basis of organic patterns, *On Growth and Form* (1917) (Ogawa et al. 1996, 1-7).

In light of these developments, Terada's seemingly homely study of "things Japanese" after flirtation with the frontlines of physics inquiry in the 1910s is not to be understood as a move to deny the universality of science, repudiate the advice of his mentor Natsume Sōseki and affirm Japanese particularity. Rather, its strongest significance is in a turn to the phenomena of everyday life and experience in which to find the problems that will animate the future inquiry of physics. Where the mainstream of physical inquiry in the 1920s and 1930s perceived an uninteresting muddle, Terada perceived the deepest connections of physical phenomena at the level of form, pattern and structure. In the context of a first-rate mind marginalized by virtue of his situation from the mainstream of early twentieth-century physics, this is a clear, strategic move to locate the mystery and uncertainty of the natural world not in the realm of the atom and distant space, but in the still unknown problems of classical physics, the regularities that emerge from the non-linear, non-continuously differentiable phenomena of everyday life. But the claim that Terada had grasped ahead of his time the essential problems of complexity, the forms it would take, and had some glimmering of the field 50 years before it formed, can best be judged with reference to his own reflections on the problem in 1931, from an essay called "The Problems of Physics Around Us In Daily Life," which merit quotation at length:

It is little exaggeration to say that contemporary physics is close to powerless when it comes to the statistical laws governing something like the forming of condensate into drops of water and their subsequent flow down a vertical surface. Research on this kind of problem is scattered to the winds. For example, between the article on the distribution pattern of "fractures" by Taguchi Ry∈zaburō in the current issue of this magazine [*Kagaku*, Apr 1931], and such problems as the curious form of the image in Lichtenburg's electric discharge [リヒテンベルク放電像], the breaking down of a drop of fluid as it falls, the development of spikes and corners in a sugar-plum as it grows, the sudden occurrence of a slippage in a simple metallic crystal [金属単晶の滑り面], and, from a slightly different perspective for example the mode of bifurcation in a river, the distribution of branches in a tree, the growth of a stripe pattern in a clamshell, even in this kind of *extremely complex* problem it is possible to perceive a certain "principle of form" stretching out at the most fundamental level and unifying these problems. That is to say, all of these problems are bound up in the question of "stability versus instability." In most cases where instability enters the problem, the state of affairs becomes statistical, and tends to be excluded from physics as configured up to now. Typically cases like this are dismissed without a second thought under the pretext that, "the phenomenon is not reproducible." And indeed, the term "reproducible" as it has been used to the present day has meant reproducible in the deterministic sense, hence that sort of response seems entirely natural, however even in the case of so-called "unreproducible" phenomena like those outlined above, in a statistical sense they can be reproduced with the greatest precision. In this way "laws" for these phenomena exist perfectly well in a statistical sense. Really, though, even in the case of determinism as it is usually understood in physics, once the inquiry leaves the domain of sense-perception and enters the subatomic world, the problem again comes undone into the realm of statistics, probabilistic averages and their variations. And it appears at the present day that there are going to be placed untranscendable limits even to this statistical knowledge. In an age like this, if you ask just what kind of knowledge we do possess in the field of statistical physics, it depends on one's perspective, but I would have to say it is very poor. The laws of classical statistics have been shaken by the findings of nuclear physics, and even if a new kind of statistics is developed to take its place, the question of *the relation between these two statistics*, each valid in their own domains, becomes very difficult to conceive. Setting that aside for the moment, even with the problem of how to come to an understanding of how the water droplets gather into a stream on the glass door before my eyes, even if I exhaust the literature on the subject there is hardly even a place for me to gain a foothold. Is that really because I'm seeking something where nothing lies? That's obviously not the case, the phenomenon exists before my eyes. And phenomena do not occur for no reason at all. We simply haven't been able to grasp, describe and discover the laws behind it yet. (Terada 1991, v. 3, 268-269, italics added)

Waiting in tense expectation for that unpredictable moment when a droplet of water beads, and runs down the plate of glass, and unable to shake these reflections about the tendency of physics to declare off-limits problems it cannot handle with existing protocols, Terada is unquestionably on the territory of complexity. Richard Feynman muses uncharacteristically in his 1961 Lectures on Physics that while the task ahead for the foreseeable future is mainly to trace out the reductive paths already posed by current research, the next great awakening of human intellect may involve a method of understanding the phenomenal contents of the fundamental equations of physics.[6] In an interview from 1979, he turns this reflection to the most difficult problem of all, the turbulence of fluids:

> Look at the equations for the atomic and molecular forces in water, and you can't see the way water behaves; you can't see turbulence. . . . With turbulence, it's not just a case of physical theory being able to handle only simple cases—we can't do any. We have no good fundamental theory at all. (Feynman 1979)

People in fields that deal with turbulence—meteorologists, oceanographers, geologists, airplane designers—are referred to as being "up the creek" because of this lack of fundamental physical equations. Feynman adds, "[a]nd it might be one of those up-the-creek people who'll get so frustrated he'll figure it out."

---

6. In this he echoes Heisenberg in his discussion of the reductive path of science in *Philosophical Problems of Nuclear Physics* (1949).

This quality of being "up-the-creek" is probably as good a definition of complexity as is available today. Terada was one of those up-the-creek people, eschewing the deep mainstream to search for the fundamental principles behind the wealth of phenomena right before our eyes.

Reductive and Non-Reductive Thinking

If Terada's posthumous reputation has rested mainly on his identity as an essayist and writer of lyric miscellany, the renewed attention being given his scientific work (Arima 1995; 1996; Ogawa et al. 1996; 2000) forces a reevaluation of his split career. To consolidate this impression, let us look at Kawakami Shinichi, a geophysicist with the Nagoya University Earth Sciences Group, and one of the principle figures in a new program called "Stratology" (*shimajimagaku, kôkôgaku*), an offshoot of geology and plate tectonics that sees the next century of earth and planetary sciences engaged in a full-scale, interdisciplinary effort to interpret the volumes of data embedded in the layers of sedimentary rock in the earth (Kumazawa and Itō 1996). Kawakami chooses as his epigraph for an article in which he glosses the project as a "comprehensive hermeneutics of earth history" (*zen chikyû shi kaidoku*) a quotation from an essay by Terada written shortly before his death called "Stripe Patterns in Nature."

> In the event the appropriate "teeth" were found to dig into and fully digest the significance of these phenomena . . . that is to say, an appropriate methodology . . . we might very well see them taking center stage as one of the fundamental problems animating the scientific world. (Terada Torahiko, *Kagaku*, 1934)

Kawakami goes on to bring the problem up to date:

> Stripe patterns are found throughout the natural world. I think one can say, though, that Terada Torahiko was one of the first people to really discern the importance of the meanings they hold. Terada's vision is now being given concrete shape in the field of earth sciences he loved so well under the keyword "stratology." It is becoming understood of late that in stripe patterns such as the rings of a tree, or the strata of the earth is inscribed a history of the movement of the earth in the universe. And it is in Japan where systematic research into this has first begun. (Kawakami 1996, 210)

Natural scientists in these speculative areas, stereology, the science of form, stratology, share a desire to mobilize simple, non-analytical models for knowledge (the stripe, the form) as a way to shock their disciplines out of an isolating, tree-like model of knowledge that is limiting their ability at present to make interdisciplinary connections among the enormous fields of data being produced (the relation with computers is unshakable). And they share a recognition of Terada as a forerunner of their field. Norbert Wiener, early information theorist and originator of the term cybernetics, defines a message as "a pattern distributed in time" (Wiener 1950, 21). Terada realized before the advent of computers that in stripes, in the accretion of layers, and the resulting patterns was information, and this information was shared across a variety of natural processes. This is the insight from which these speculative new sciences proceed, the sharing of data across fields.

The fact that "stratology" was recognized by the Ministry of Education in 1995 as a new "field of emphasis for research" (*Monbushō jūten ryōiki kenkyū*) recalls a point raised by Tessa Morris-Suzuki in an article in the volume *Japanese Encounters with Postmodernity*, where she wisely counsels a skepticism toward the idea that certain "postmodern" fields of science and technology are a more direct expression of some set of Japanese or Asian cultural traits than the reductive, analytic and instrumental mainstream of science. And in the irreverent writings of the members of the Nagoya Earth Sciences Group (who describe their laboratory as a "salon") it is clear that a certain irony is intended and they themselves are not sure how seriously to take their own stratology project and the interest it has generated. Yet if one thing is clear in the history of science, it is that truly original science work contains an element of play and a simple curiosity in new configurations, and new ideas tend to be regarded as jokes, as unserious when first floated in a scientific community governed by different paradigms (Kumazawa and Itō 1996, 124; Valdés and Guyon 1998, 28-29). Hence, while one must accept Morris-Suzuki's elucidation of the compatibility of concepts like "fuzzy logic" and "holism" with the set of ideological notions used to underpin a notion of Japanese postmodernity, that doesn't mean they aren't right, and such a caution in the face of speculation can just as easily translate into the equally conservative ideology that the particular course science has taken to the present in its engagement with the manifold of natural phenomena will determine the future course of science. Hence, the same skepticism and caution needs to be exercised about the idea of a new science of

Form based on the more loose and capacious term "katachi," or of a stratology based on hermeneutic rather than analytic principles. Yes, they posit a unique suitability of Japanese scientists based on cultural, linguistic and institutional conditioning to map and advance these new directions, and yes, their validity still needs to be judged, as Kaneko hints, on the same questions of economy, predictive value, and ability to suggest new lines of research as any other field.

The term "hermeneutics" is not introduced unadvisedly here, and refers to specific institutional and philosophical points at issue for the proponents of complexity as a research program in Japan. The standard way of distinguishing the two major traditions of modern philosophy is in terms of their historical lineage, "tracing each back, through G. W. F. Hegel and Gottlob Frege respectively, to their initial point of divergence in the nineteenth-century inheritance of the Kantian legacy." If this has produced a division in the humanities in North America between an Anglo-American or "analytic" tradition which conceives of the world as a series of problems, each in search of a solution, and a continental, or hermeneutic, tradition which conceives of the world as a series of texts, each in search of a reading (Conant 1991, 616-617), both traditions are well-represented in scientific thinking in the difference between experimental and observational sciences. In observational sciences like cosmology, evolution, plate tectonics, etc., theories are not subject to laboratory experiments, but are evaluated by their explanatory power, that is to say, their capacity to account for the totality of the facts in a coherent way (Zaldarriaga and Hogg 2000, 2079-80).

This difference is well-marked in Japanese discourse, such that the discussion of complexity by scientists takes on a specifically hermeneutic tone. According to Kaneko Kunihiko, a physicist at Tokyo University and recognized expert in chaos theory:

> There is a tendency to confuse the new research on Complexity (*fukuzatsukei*) with work on systems that are complicated, but these need to be differentiated in terms of their research orientation. In the latter case, one can find all sorts of factors intertwined in a problem, but in the end, if these can be disentangled and analyzed separately, some sort of understanding will result, and in this sense the complexity is only apparent. Research in Complex systems begins when one faces a problem in which this kind of analysis of factors still doesn't explain the phenomenon. In other words, rather than combining the solution of parts to yield an understanding of the whole, one enters a circular process (*junkan*)

whereby it is only through a grasp of the whole that you can understand the individual parts. (Kaneko 1996, 79)

This of course is a description of the hermeneutic circle. Elsewhere scientists use the language of hermeneutics when they speak of what it means to "know" the world, of the need to have a totality in mind to model processes, and of the problem of complexity inhering in the resulting possibility of a "context-level causality" removed from the observed system (Yoshioka 1996, 190; Matsuno and Tsuda 1996, 53). This type of language produces a sense of recognition in the literary critic, and one might be tempted to feel that science is finally coming around to the insights that pass for common place in theory and to want to help the scientist, saying, "well if that's all it is, you should have come to us sooner, we could have told you that a long time ago." Scientists refer to this as armchair philosophizing (Feynman 1997, 73-77). Kaneko continues, though, and here we see the standard on which Terada's contribution to Complexity will be judged: "However, if it's just a matter of saying, 'reductionism has reached a dead-end,' anyone can do that. The question is whether one can stand up a program of scientific research from that" (Kaneko 1996, 79).

The Maverick Precursors to Complexity Theory

Given the fact of a renewed interest in Terada's work among a certain avant-garde of scientists, we are now in a position to stand up an if-then proposition: If Terada is, as claimed by a number of natural scientists in Japan and argued here, really a precursor to complexity, a figure who sounded the field before it was comprehensible to the reigning paradigm, then he needs to be grouped with the other early twentieth-century figures that pioneered the field. We may turn to Benoit Mandelbrot to find a gallery of European contemporaries of Terada who sacrificed their careers trying to stand up such programs.

In his 1977 book *The Fractal Geometry of Nature*, Mandelbrot, a researcher at IBM, gives an essentially mathematical presentation of a peculiar entity: recursively defined curves and shapes with statistical regularities, whose dimensionality is not a whole number, and relates these to the description of the natural world. Though Mandelbrot had been groping toward the idea of the fractal in a series of obscure papers dating from 1951, the systematic presentation of these ideas in 1977 was made possible by the advent of computers, and all illustrations of dusts, trees, islands, and snowflakes are generated by computer programs. The presentation is throughout spare and mathematical, however, Mandelbrot includes a section at the end with biographies of a number of scientists and mathematicians from the early twentieth century, who prefigured the field before computers were available (Mandelbrot 1983, 391-424).

If you have spent any amount of time with scientists or science writing, this will strike you as idiosyncratic, because scientists seldom get into the details of biography beyond birth and death, in fact the irrelevance of biographical information on the author of a paper is almost a point of professional pride. In the humanities, by contrast, even outside the genre of biography, devoting substantial proportions of an argument to discussion of the personal details of a thinker with whom one is concerned is de rigeur, a custom untouched in practice even in the halcyon days of deconstruction and the death of the subject. Where was the person born, what was their schooling like, when did their mother die, etc. We know these things. We know that Sōseki was given up for adoption as a child, then returned. We know that Miyamoto Yuriko was the daughter of a wealthy, bourgeois family, that Mishima Yukio had a controlling grandmother, and that Dazai Osamu drank a lot. We know that Terada was a member of Sōseki's coterie. It has never been clear to me why one feels compelled to discuss these details, and one cannot overestimate the importance of this point in differentiating the humanities and the sciences.

However, Mandelbrot himself went through a long period of groping before arriving at the order toward which his apparently disordered career had been leading, and he reserves a special place in the case of one type of scientist: the maverick. Though most scientists' lives are uneventful with the exception of the occasionally colorful circumstances attending the revelation of their talent, and "a life story that is interesting to tell is rarely the reward of those who keep to the mainstream of science" (391), there is one particular type destined, or cursed as the proverb would have it, to lead interesting lives. For these scientists, the biography has a place in the story.

> Tales according to the typical pattern are legion, but the stories selected for this chapter are entirely different. . . . The hero is a loner. Like certain painters, he might be called a *naive* or a *visionary*, but there is a better term in American English: *maverick*. When the curtain falls on the prologue of his life story, he is still, by choice or by chance, unbranded. (392)

Mandelbrot numbers among the forerunners to fractals and complexity theory seven mathematicians and scientists from the early twentieth century: Louis Bachelier (1870-1946), D'Arcy Wentworth Thompson (1860-1948), Edmund Fournier D'Albe (1868-1933), Harold Edwin Hurst (1880-1978), Paul Lévy (1886-1971), Lewis Fry Richardson (1881-1953), and George K. Zipf (1902-1950). Here are the characteristics he finds common to them.

First, their careers tended to be lackluster and entry into the mainstream is postponed. They are perennially unsuccessful applicants, fail to achieve the recognition of their peers, and work outside the centers of their field. Second, they tend to have a certain amount of leisure, by choice, or enforced by the circumstances of their career. D'Arcy Thompson's daughter writes that it is "a matter of speculation whether [his magnum opus *On Growth and Form* (1917)] would ever have been written if its author had not spent thirty years of his early life in the wilderness." Few advance to key posts in major research institutions, their administrative duties are few. "The key," according to Mandelbrot, "seemed to be time to spare." Third, they have a certain kind of strong mind that does not relinquish their stance despite lack of encouragement from their peers, and tend to do their best work very late. D'Arcy Thompson published *On Growth and Form* when he was 57, and Levy's major books were written at 50 and 60. "The cliché that science is a very young man's game is definitely not true in their case" (392).

And fourth, it is a common fate for their reputation to skip a generation. Mandelbrot writes of Louis Bachelier, whose books and papers went unread in his lifetime, but whose work on probability theory is now recognized as seminal in math and economics, that his tragedy was "to be a man of the past and of the future, but not of the present" (394). Lewis Fry Richardson, whose work on weather prediction by numerical process was viewed as disreputable for twenty years after publication, was reprinted after 33 years as a classic (402). And we may add Wegener's theory of continental drift, which had to wait well into the 1970s to be accepted. A final, contingent historical factor is that these thinkers proceeded on intuition before computers were available to handle the problems posed by their research.

Terada shares all these characteristics fully. He shares the problem consciousness of these pioneers, and fits the narrative, right down to the reputation skipping a generation. Hence we must inquire into why it did not occur to Mandelbrot to include Terada in their company.

Doing the Math

In reading Mandelbrot, it is clear that his benchmark in judging contributions, in distinguishing figures who really set their mark and deserve true fame and those who merely deserve footnotes, is a lasting mathematical result, and it is on this point that Terada's achievement rises or falls. In dealing with the fact of the endorsement of Terada by Japanese physicists, one feels a certain uneasiness with a vagueness about the failure to push to mathematical expression, both in the work on the Laue phenomenon, and in his subsequent work on patterns and forms. If the mastery of classical mathematics is still the benchmark of a real contribution, this is a damning omission.[7]

Perhaps the closest comparison to Terada is the naturalist D'Arcy Thompson. Leafing through the now classic 1917 book *On Growth and Form*, it is immediately clear that Thompson is involved in basically the same kind of problem as Terada, discerning principles and connections among the seemingly disparate *patterns* that appear in the natural world. D'Arcy Thompson, however, restricts his investigation to organic forms. But his strongest move is in reaching out to physical and mathematical principles to elucidate those forms.

> The terms Growth and Form, which make up the title of this book, are to be understood, as I need hardly say, in their relation to the study of organisms. We want to see how, in some cases at least, the forms of living things, and of the parts of living things, can be explained by physical considerations, and to realize that in general no organic form exists save such as are in conformity with physical and mathematical laws. (Thompson 1992, 15)

Life can be distinguished from inorganic matter such as stones, wind, planets and rain by three properties: living things can grow; they can reproduce; and they can react to changes in their environment. All living things possess all three characteristics. Newton laid down in the *Principia* the mathematical principles governing the movement of *inorganic* forms, the density and resistance of bodies, the motion of light and sounds, and spaces void of all bodies, as a basis such "as we may build our reasoning upon in philosophical inquiries" (Newton

---

7. I am indebted in understanding Terada's position in the history of science, to the response by Sidney Perkowitz, Candler Professor of Physics at Emory University, and chief consultant in the American Physical Society's "Century of Physics" project, to an early version of this paper given at the Southern Japan Seminar, Panama City, FL, 24 September, 1999.

1952, 269, hereafter Principia). The "mystery of life," then, posed in its canonical form by Helmholtz and quoted by Thompson, is whether there is something *else* going on in organic forms, whether "there may be other agents acting in the living body than those agents which act in the inorganic world" (12). In setting up his magnum opus, Thompson goes against the consensus of his day, and denies that there is a special set of laws governing the generation of forms in living matter, and reaches out to physics for their principles of propagation.

Though Thompson focuses rigorously on the question of organic phenomena, he is well aware that the largest stakes in his denial of a special set of laws to the physical forms of life, and application of physical and mathematical principles to the process of growth, lay in the production of a theoretical medium in which the generation of organic and inorganic forms would be revealed to share the same physical principles. That is to say, implicit in his project is that he sought to produce a science of form that would encompass the patterns of nature: both organic and inorganic.

The reason why this is implicit is contained in the "Rules of Reasoning in Philosophy" by which Newton precedes Book III of the *Principia*, The System of the World. For Rule II, Newton gives:

> *Rule II: Therefore to the same natural effects we must, as far as possible, assign the same causes.*
>
> As to respiration in a man and in a beast; the descent of stones in Europe and in America; the light of our culinary fire and of the sun; the reflection of light in the earth, and in the planets. (270)

The insight which governs Terada's later work follows exactly this line of reasoning. He does not duplicate D'Arcy Thompson's work, he follows classical scientific reasoning, and takes his next step. Versus the trees, leaves, horns, cellular patterns, seashells and teeth and bones one finds investigated in Thompson's organic domain, one finds in Terada's work the propagation of waves, bifurcation of lightning, rivers, spark discharge patterns, the fracture patterns in concrete, glass, etc. The project of bridging the organic and inorganic in a science of form becomes explicit in the late experimentation on patterning in animal fur. Terada sits and contemplates the patterns of black and white on a domestic cat, a living body, and through a series of practical mapping exercises, finds that the pattern on the elliptoid body of the cat is a projection of the way fracture patterns propagate on the flat, inorganic surface of concrete, and extends the analysis to the pattern of division of egg cells. Terada concludes that some common principle must underlie their forms.

D'Arcy Thompson does touch on a concrete example of inorganic pattern formation in the revised 1942 edition, where one finds a photograph of a porcelain bowl from Japan. The example is taken from a paper called "Cracks Upon the Glazed Surface of Ceramic Wares," published in the *Scientific Papers of the Institute of Chemical and Physical Research of Tokyo*, 1935, by an H. Hukusima (See Fig. 8), and Thompson also cites "On the Origin of Colored Patches of Some Kidney-Beans," by an M. Hirata published the following year in the same journal (Thompson 1992, 522-525). Fukushima Hiroshi and Hirata Morizō are Terada's students, a bit of serendipity that hints at the extraordinary affinity in the projects of these exact contemporaries, Terada and D'Arcy Thompson, and at a collaboration that might have been. You will recall the Institute for Chemical and Physical Research in Tokyo as the site of Kangetsu's fictional paper on the physics of hanging in 1905.

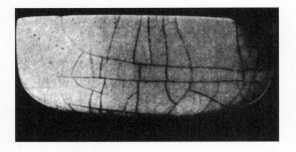

Fig. 8 Illustration from H. Hukusima, 1942 Edition

What secures Thompson's place in the history of complexity, though, ironically in that Terada is the physicist, is his resolute and systematic concentration on placing his intuitions on a mathematical footing. One cannot underestimate the difficulty of the task, and this by no means indicates a full elaboration of the problem. Perusal of the 1000 pages of *On Growth and Form* yields only the most basic math, going nowhere beyond the basic trigonometry and calculus one would find in Clifford's famous handbook, *Common Sense of the*

Exact Sciences (1887). D'Arcy Thompson clearly admits this in the preface:

> This book of mine has little need of preface, for indeed it is 'all preface' from beginning to end. I have written it as an easy introduction to the study of organic forms, by methods which are the common-places of physical sciences. . . .
>
> It is not the biologist with an inkling of mathematics, but the skilled and learned mathematician who must ultimately deal with such problems as are sketched and adumbrated here. I pretend to no mathematical skill, but I have made what use I could of what tools I had; I have dealt with simple cases, and the mathematical methods which I have introduced are of the simplest and easiest kind. (unpaginated Prefatory Note)

Thompson never gets close in *Growth and Form* to the "monstrous" mathematics of fractal geometry, and Mandelbrot undoubtedly saw his place prepared in the passage above. The point of the role of math in differentiating unfettered speculation and hypotheses from science is not in sheer mathematical sophistication, it is rather in the feel for the way the domains and relations of math are mapped onto the physical world. It is for this "easy introduction" of math into a science of biological form that till then had more closely resembled aesthetics and history, that D'Arcy Thompson is given credit for a lasting result. In Kaneko's words, Thompson is recognized today because he "stood up a program of research" from his intuition. Terada did not or could not, and it cost him his place in history. This was the problem with the Laue phenomenon work in 1912, and this is what denies him his place in Mandelbrot's pantheon. That is why his superior intuition cannot be recognized, at least so far, at least outside Japan.

Sōseki has another recurring scene in *I Am a Cat*, where the protagonist Kushami goes around boasting of his indigestion after he finds out his intellectual hero Thomas Carlyle was also dyspeptic. His friend Meitei, obliging as always, points out the fallacy: "Carlyle was indeed a dyspeptic, but it does not follow that all dyspeptics are Carlyles." Terada shares a remarkable array of the paradoxical qualities of the early pioneers of complexity, but it does not follow that he pushed it to a lasting result.

Resolving the Question of Creativity

The quality of Terada's intuition is not in doubt. One has to take seriously though the possibility that the investment of time and mental energy to the writing of 30 volumes of literary miscellany might have come at the expense of his physics.

Both Terada and D'Arcy Thompson were leisurely, well-read, philosophically inclined scientists removed from the pressure to conform of the mainstream. Both used the contemplative distance afforded to search for connections and principles among the seemingly random accumulation of patterns in nature. What does this say about their commitment to science as a mode of understanding? Rule I of Newton's "Rules of Reasoning in Philosophy" asserts that one is "to admit no more causes of natural things than such as are both true and sufficient to explain their appearances." The rule goes on to add, "Nature is pleased with simplicity" (*Principia*, Book III). This provides a key to what is different about the science of Terada and D'Arcy Thompson, and what set them off from the mainstream. The intuitive act of making connections among seemingly disparate things is classical scientific reasoning. However the relation to Newton's prescription is ambiguous: Terada and Thompson are reductive, yet find nature pleased with complexity. They seek "the same cause for the same natural effects," however come up for the geometry of nature not with conic sections, neat geometric progressions and the inverse square principle, but with the pathological forms of complexity. The precursors of complexity, then, were poised on the border of a different kind of thinking, which, recalling the rare balance of Terada's oeuvre, and exploiting an anomaly in current neurophysiological models of creativity, I will call "literary." Let us try to specify.[8]

Neuroscience as a field is concerned to find causal, physiological factors in the brain for higher-level mental processes and capabilities, such that one can make a statement of the form: If physiological factor A is present, then we will have higher-level phenomenon X. If you cannot make a statement of that form, neuroscientists just don't care, you are outside of the domain.

---

8. I learned a great deal by participating in the consistently stimulating weekly CNS (Cognitive Neuroscience/Central Nervous System) seminars held by the Department of Neurology in the University of Florida College of Medicine between 1999 and 2003. Research presentations on rotating topics followed by question and answer were instrumental in forming my understanding of the approach and methodologies of research on the physiological basis of behavior.

One of the principal methodologies of neurophysiology is, when investigating any particular cognitive process or capability, to identify a population which has either a) lost a universal capability, or b) exhibits an unusual trait in a high degree, and then search for some sort of physiological difference that accompanies these extreme states. By slowly drawing threads between cases, the researcher hopes to close in on a causal, neurophysiological explanation for the phenomenon. Vis-a-vis a universal or nearly universal skill this most often means identifying a population with a *dysfunction*, often through injury or stroke, as for example with motor skills (cannot walk, point, grasp); language skills (cannot use analogies, remember words, name an object in one's left hand); reading skills (reverses letters, cannot switch between speaking and reading, etc.); or American citizen skills (are not happy and optimistic, cannot concentrate at school, and so forth), and seeing what is different about their brains in terms of its physical properties, and chemical and electrical processes. This is why patients with localized brain trauma are prominently featured in neuroscientific research: they provide a clear starting point for drawing the strands together. When strong correlations are found between damage to parts of the brain and loss of particular cognitive skills, this knowledge becomes instrumental in that it allows intervention by doctors in the form of drugs, surgery, various forms of therapy, and modification of early childhood practices in order to return these dysfunctional populations to the universal group, to allow the stroke patient to speak and use their hands again, to make the depressed person happy, energetic and optimistic, to make the disruptive child tractable and obedient. The normative thrust of this research is obvious, and everywhere assumed with confidence.

However, when the trait is something that is distributed scarcely in the general population, such as creativity and the related concept of genius, researchers face the slightly different task of identifying a population that has it in a high degree.

Compared with the systematicity, confidence and sense of mission which marks the growing neuroscientific literature on pain, motor functions, the emotions, schizophrenia, intelligence, depression, language, etc., the literature on creativity is characterized by an atypical tentativeness. There is first the problem that the effort to define creativity by nature exposes the research community to the charge of ridiculousness Pascal sees leveled at those who wish to treat matters of intuition scientifically. Second, although there is a circularity inherent in the project of neuroscience as a whole (the trait to be isolated and defined is assumed in identifying the initial sample), the problem is particularly acute in the study of a relatively poorly delineated concept like creative genius, where the researcher exerts proportionally more influence. The process of selecting a sample of creative people runs the risk of simply confirming the investigator's prejudices about art and culture, or of relying on social conventions that are historically and culturally specific. Additionally, the relation of genius and insanity is overdetermined in the Western tradition,[9] which makes it difficult to disentangle cause and effect when one considers that audience and critical expectations, as well as artistic practice, are conditioned by romantic stereotypes of genius. Given the demonstrable rise and fall of the "stock" of artists, musicians and writers over time, the step of identifying highly creative people has been criticized as equivalent to the purchase of equity in the stock market, whose valuations are based not on any inherent quality of the companies, but on the expectations, perceptions and desires of the investing, or investigating community (Holland 2000).

In order for the first tentative threads to be drawn, neuroscientific studies of creativity tend to take as an initial sample figures from the past whom the passage of time has ratified. Beyond such obvious traits as drive, perseverance, love of novelty and laxity of repression, a recurrent association has emerged between individuals of high creative achievement and the presence of psychopathologies, such as bipolar disorder, depression, schizophrenia, alcoholism, and so forth. 16 out of 28 articles in a recent special issue on creativity of the *Bulletin of Psychology and the Arts* (1:2, 2000) feature the word psychopathology, madness or schizophrenia in the title, a typical formulation being, "Are there mental costs to creativity?" The association of genius and insanity is a cliche from the romantic tradition. However, framing the question in terms of creativity and psychopathology has allowed a point to emerge that brings up the particular concerns of cognitive neuroscience: a correlation of creative achievement with low cortical arousal.

---

9. The idea goes back to antiquity, as in Plato where the poet or artist is seen as in the grip of divine inspiration, and receives its canonical formulation with the glorification of genius and the irrational in the Romantics, taking an unfortunate detour through phrenology and Lamarckian adaptation in Lombroso in the late nineteenth-century (Martindale 2000a, 28-30; Lombroso 1895). This has an imperfect counterpart in the figure of the religious shaman, and the wise man as fool (*gu*) in Chinese and Japanese traditions.

If one considers the canonical scenes of creativity, they are strikingly similar in picturing a state of reduced mental tension. There is Newton under the apple tree, Einstein in the patent office, Archimedes in the bath. Shakyamuni wanders for years, then sits down under the Boddhi tree, and is enlightened in an instant. The self-reports of creative people indicate that inspiration occurs in a reverie-like state accompanying moments of repose and defocused attention as one sinks into a relaxed state. According to Heilman, there are three main "scenes" for such changes in creativity: First, changes in creativity are known to occur around sleep, often in the period of twilight or relaxation before sleep, or in dreams. Ideas visit one fully formed, and one leaps out of bed. Second, they occur during relaxation after a period of tension, as in the aristocrat retiring to the country.[10]

Creative achievement also seems to cluster, however, around a set of pathologies, and one need not enumerate the many examples of composers, artists and writers who suffered from bipolar disorder, depression, alcoholism and drug addiction. Conversely, studies have also consistently found in the family histories of people at risk for schizophrenia a higher than expected number of creative people, as well as criminals, alcoholics, and people with affective disorders (Heilman 2000; Eysenck 1995; Martindale 2000).

These would seem to exist at the level of cliche: the canonical stories and the genius-madness nexus. However, as it turns out, there is a neurophysiological factor that unites them that gives the seemingly odd set of correlations a certain robustness as an explanatory strategy: namely, all are characterized by low central nervous system levels of arousal. Here the threads begin to be drawn together. The theoretical connection of creativity and low cortical arousal began with Hull's *Principles of Behavior* (1934). Coming at the problem from a behaviorist point of view, i.e., leaving the physical brain a black box, Hull observed that increases in arousal (which he refers to as drive) "make the dominant response to a stimulus even more dominant" (Eysenck 1995, 267). That is to say, stress, pressure and high states of arousal decrease the tendency of the brain to sort through the marginal and low-probability connections that by definition cause creative thinking

---

10. These points are confirmed by experience in such low-level, "normal" acts of creativity as preparing a lecture, planning an elaborate meal, or writing a book on science and literature, where puzzles intractable during the working day sometimes unravel spectacularly at the moment of surrender on entering a shower or collapsing into bed, yielding exactly the solution for which one has been looking.

to "stand out." Martindale (1981) argues that, when analyzed, "a creative idea is always one that brings together . . . seemingly unrelated ideas," and explains the relation between depressed states and this model of creativity in terms of a "defocused attention."

> Attention is the focal point of one's conscious experience. Narrowly focused attention is accompanied by a high level of cortical arousal. As cortical arousal declines, attention becomes less and less focused.[11] (Martindale 2000, 50)

Subsequent studies have elaborated this observation of an inverse relation between stress, anxiety and other states of arousal (triggered by noise, the presence of other people, insistent demands on time) and creativity in terms of "attentional narrowing," "cognitive inhibition," and "steep associational hierarchies." That is to say, states of increased arousal direct the brain to more salient schema that have proved effective through past experience. The relaxed cognitive states associated with creativity, then, are not simply characterized by an indifference or a slackness, but an active negation of these states, hence they have names like "cognitive disinhibition," "defocused attention," or "broad associational hierarchies." That these are functional versions of more extreme states associated with psychopathology will not escape attention (Martindale 1981; Eysenck 1995).

In drawing the threads together between this behavioral and historical data, however, neurophysiological accounts must seek a causal, physiological explanation. The attention and arousal levels found to correlate with creative thinking are known to be under the control of cortical neurotransmitter systems, and share a nexus of characteristics with depressed, bipolar, and schizophrenic states (Martindale 2000b, 50). Heilman suggests that decreased norepinephrine seems to depress established neural pathways and lead to heightened cognitive flexibility. Tension, stress and pressure, on the other hand, while increasing arousal rates also increase the probability that answers will be arrived at according to established pathways. Eysenck, on the other hand, argues that high levels of dopamine and low levels of serotonin cause the defocused attention found in creative individuals. Both seek to explain the strange correlation between creative achievement and a variety of depressive or schizophrenic states by positing

---

11. This recalls the continued validity of Sōseki's wave model of conscious experience. The wave broadens out at moments of reverie, facilitating association.

some intersection of physiological traits that lead to the low cortical arousal common to each.

Other conditions being in place, then, this at-first-sight counterintuitive relation of depressed cortical activity and creativity seems to have withstood elaboration in terms of neuroscience, and demonstrated explanatory power in terms of other independent psychological variables. In "Serendipity in Poetry and Physics," Valdes and Guyon speak of creativity as "a playful looking for understanding or curiosity in new configurations of meaning, a playing a game of 'as if' or 'what if' with whatever we catch in our net of imaginative reflection," and call doing physics and reading poetry "two well-known and long-established playful activities" (28-29). The location of physics and poetry in opposition to work, which seems a provocation at first, in fact has a basis in the neurophysiology of creativity.[12] The reason this can be applied in a speculative way to Terada is because, in the context of the desultory emergence of the idea of complexity in scattered and isolated researchers in the early twentieth century, it transforms the thread Mandelbrot identifies as common to these forerunners of the various fields of complexity theory, namely a certain leisure, enforced or unenforced, due to their unsuccessful careers vis a vis the mainstream, into an explanatory factor. Given a "peculiar kind of strong mind," "the key seems to be time to spare" (Mandelbrot 1976, 392). Given a first-class physics mind, the leisurely pace enforced by Japan's distance from the high-pressure centers of research in Europe, the relatively relaxed culture of Tokyo University versus the more aggressive physics departments at Tōhoku and Kyoto, and Terada's year-long convalescence in 1919 after collapsing from illness become enabling conditions for Terada's far-sighted work. We have no need to be *surprised* to find ground-breaking work in the quiet physics labs of Tokyo University in the 1910s, because the point about creativity steers one to assign a positive significance to the lack of focus in Terada's research, and to link these analytically with his status as a naive or visionary vis a vis complexity. Terada was busy making low-probability connections.

A lesser known part of Newton's story is that the reason he was sitting under an apple tree in 1665 is that his post had been dismissed

---

12. Valdes and Guyon (1998) spell out the professional competence that has to be in place for moments of serendipity to have meaning within a discipline, while Martindale (2000b) summarizes the list of more or less common subtraits that have to cohere in a single individual for the rare "emergenic" trait of high creativity to occur.

at Cambridge for two years due to the spread of the Great Plague from London. During this forced retirement he discovered the theory of gravity, invented the calculus, and discerned the composite nature of white light, a series of feats rivalled only by Einstein in the patent office. The scientists and mathematicians who sounded the problem of complexity in the first decades of the twentieth century had these necessary, if not sufficient conditions to do path-breaking work. However, they ended up like Terada less well integrated into the story of science, an interesting and lonely group who failed in most cases to integrate their insights into any predominant directions of the field.

Reductive and Non-Reductive Creativity

This brings up an interesting anomaly in the neuroscientific literature on creativity, a divergence between scientists, and artists and writers on one point. Recalling the canonical scenes of inspiration from the previous discussion, the defocused state that enhances creativity seems particularly likely to occur under three conditions: First, in the moments surrounding the passage into sleep, second in the moment of relaxation after a period of tension, and third, in correlation with certain pathological states, such as bipolar disorder, depression, alcoholism, etc. Scientists and artists/writers seem to share conditions one and two as scenes of inspiration, however a discrepancy that is not well-handled by scientific research into creativity is that they diverge on condition three. The incidence of psychopathology or mood disorders in scientists is no greater than the general population.

> Here is the basic paradox. We know that eminent members within the creative arts have a far higher lifetime prevalence of mental illness than those in the sciences. We also know that creative artists and scientists show no appreciable differences in the extent of their originality, innovativeness and creative achievement. So within the creative arts, there appears to be a relationship between mental illness and the extent of creative achievement, but within the investigative professions, creative achievement tends to be associated with mental stability. How do we explain these discrepancies? (Ludwig 2000)[13]

---

13. Ludwig begs the case by concluding that scientific work is logical, objective and formal, while artistic work is "intuitive, subjective and emotive" (48). This is a good example of the tendency of the literature to confirm common-sense notions in a circular manner.

Indeed, the great artist and the scientific genius vie in the popular imagination as the canonical instantiations of creative inspiration, hence the failure of these two prototypical samples to fall out neatly must indicate a problem with the framework organizing the data. None of this is to suggest that all artists are bipolar, or that no scientists are depressive, these are statistical correlations. However, when an anomaly consistently appears in the data, this indicates not a problem with the data, but with the concepts by which it is being organized. I would suggest that this persistent anomaly in the data can be solved by introducing a heterogeneity within the idea of creativity itself.

The acts of association and curiosity in novel configurations that define the creative act are glossed by Heilman as "the ability to find unity in what appears to be diversity." This seems plausible, but actually goes a step beyond the broad acts of association and juxtaposition conditioned by disinhibition and states of low cortical arousal. It is possible to imagine another response to the serendipity of a chance association, an insistence on the irreducibility of the difference. That is to say a drive to proliferating complexity, rather than simplicity. Indeed, Heilman's gloss is suspiciously close to the definition of reductionism in science, and this may reveal an implicit prejudice in the approach of researchers. Reductive explanation characterizes the experimental sciences, and involves the explanation of phenomena "wholly in terms of simpler entities." When an appropriate account of lower-level processes is given, the explanation of the higher-level phenomenon falls out (Chalmers 1996, 42-43). Richard Feynman characterizes this as "put[ting] together things which at first sight look different, with the hope that we may be able to *reduce* the number of *different* things and thereby understand them better" (Feynman, 1995, 24, italics in original).

Canonical ideas of scientific thinking always involve this search for a deep simplicity, sometimes glossed as elegance. However, it would appear that there is a parochial tendency among scientists to assume that this particular kind of intuition exhausts the category of creativity. For example, Michael Leyton's *A Generative Theory of Shape* is primarily concerned with developing a computational theory of shapes involving algebraic structures composed of group-theoretical constructs called fiber groups and wreath products. The key concepts involve the notions of transfer and recoverability, which is to say, the recognition of structures common to groups, and the ability to infer by reverse-engineering processes required to produce a given group. These concepts of transfer and recoverability are thought to grasp an essential unity between the natural world and the world of computation in the way they generate patterns. However, in reflecting on the more general stakes involved in modeling the structures of visual cognition, the author does not hesitate to venture outside his domain to the field of aesthetics:

> [A]esthetics is connected to the very foundations of our generative theory of shape—i.e., it embodies our basic principle of the maximization of transfer. . . . Transfer is the basic means by which an artist generates a work from a minimal set of elements.
>
> One of the powerful consequences of understanding that aesthetics is the maximization of transfer is that, since we will give a complete mathematical theory of transfer, we will be giving a complete mathematical theory of aesthetics. It is clear that one benefit of this theory is that it will be possible to integrate aesthetics into CAD programs in a formal and explicit way. (Leyton 2001, 33)

The correct conclusion from the premise that one has produced a complete mathematical theory of aesthetics is the computational production of art. However, Leyton shies away from this, seemingly sensing his reductive understanding has brought him to a border. This is the border with which this study is concerned, and the border identified by Pascal in the epigraph to this chapter. As with Heilman, the identification of creativity with the reductive notion of transfer is complete, and stated as a premise. However, artistic creativity, on the other hand, would plausibly seem to involve insistence on the irreducibility of individual instances and sense perception, of the particular that cannot be reduced to the general, of the non-transferable. It is the failure to perceive the possibility of non-reductive forms of creativity that introduces anomalies and gaps into the science about it. These can be resolved, at the price of the neat equation of computation and aesthetics, by delineating two types of creativity, elaborated in the sciences and the humanities, a drive toward simplicity and complexity, respectively.

Conclusion

Terada Torahiko was a professional intellectual with a rare ability to move between the worlds of literature and science, and it is difficult to name one Western scientist who maintains this level of balance. This was read by the next generation of physicists, entranced by the drama of the unfolding particle physics, to work to the detriment of his science, i.e., that it dissipated and diffused his creativity. However, the emergence of a new science has allowed a subsequent generation to argue that what looked like a skittish dilettantism was in fact a long slow approach to a problem the mainstream of physics, whose reductive success in elaborating the fundamental structure of the atom, and in uniting the different physical forces in field theory has been taken as one of the benchmarks of human creativity, could not recognize. Terada's work on the themes of complexity decades before they were practically soluble marks his mind as straddling types of creativity separated by the institution.

Newton is the only proper point of reference for judging Terada's consciousness of the problem of complexity. Because Terada was a physicist. Follow along with the *Principia*, first the "Definitions," then the statement of the smooth, stunningly simple Laws of Motion, and then the demonstration in the first three sections of Book I, on the motions of bodies. Then open your window and throw the book out. There is a breath-taking profundity to the simple second order equations that tell us it will describe a parabola in its fall to the concrete below.

The only problem is, it's not true. The book in fact meets air resistance, all sorts of vortices and turbulence and differences in pressure, the random action of trillions of molecules. Newton knows this, he is a genius. That is why he goes on in Book II to describe the motion of bodies in resisting mediums. But these are homogeneous mediums, the force of gravity uniform, and the asymptotic approach of practical mechanics and geometry is Newton's eternal assumption (*Principia*, Preface, 1). The conic sections, the geometric progressions, are an idealization. The book will fall short, in a slightly irregular way, it may be blown off course by a gust of wind, it will expand slightly from the sunlight and the heat of friction.

Engineers know this. And one of the first tasks in setting up a problem is to identify one's assumptions, to define these complications, if possible, out of the equation, and return to the idealizations of Newton. And depending on your requirements for accuracy, it is often possible, because physical phenomena approximate, under appropriate conditions and over short distances and times, the idealized, continuously differentiable behavior of the physics of point masses, vacuums and homogeneous materials. The parabola is a good approximation for most purposes. Where not possible, one resorts to statistics, there are tables, friction coefficients, safety factors.

Physics idealizes these problems, engineers wish they weren't there but deal with them as they can. However, complexity embraces these irreducible complications, the stuff and texture of life, *as the* problem. Heap up a pile of sand, now add grains directly overhead, one by one. After how many grains will the pile suddenly become unstable, and rearrange itself? This is the quintessential problem of complexity, "all bound up," as Terada discerned, "in the question of 'stability versus instability.'" (Terada 1934, quoted above) Like the phase change, like the earthquake, like water beading and running down the glass, like "the mode of bifurcation in a river, the distribution of branches in a tree, the growth of a stripe pattern in a clamshell," the direction these processes take at crucial moments lies balanced on a knife's edge. It is the reason no two fingerprints are alike, and the reason no two zebras coats are alike, and the reason lightning doesn't strike twice in the same place.

These are the problems to which Terada was consistently drawn, except for the brief interlude with X-ray crystallography. They are the weak points in classical description, all around us within the domain of experience, dismissed by a contemporary physics that could not recognize the problem.

> Typically cases like this are dismissed without a second thought . . . if I exhaust the literature on the subject there is hardly even a place for me to gain a foothold. Is that really because I'm seeking something where nothing lies? That's obviously not the case, the phenomenon exists before my eyes. And phenomena do not occur for no reason at all. (Terada 1934, quoted above)

When you line them up, with the zebras, and the lightning and the fractured glass, there is something poetic about the objects of his thought. It was only a very particular kind of person who could discern a problem here, where the physics of the day saw nothing.

One question to raise, though, is whether the strategies loosely gathered under the title of complexity can really be covered by the definition of an "irreducible particularity." Chaos, fractals, catastrophe theory, etc., despite the names, are really ways of bringing rigorous quantitative analysis to phenomena that used to have to be alluded to with names like "grainy, hydra-like, in between, pocky, ramified, seaweedy, strange, tangled, tortured, wiggly, wispy, wrinkled and the like" (Mandelbrot, 5). There is a certain aesthetic satisfaction in finding that such processes and shapes are the rule rather than the exception in nature, however the gradual introduction of complexity theory into normal science over the last 20 years, as the computing power to handle them became available, means that they will begin to be assessed according to the standard scientific criteria holding in these fields, that is to say, "mostly upon the basis of its powers of organization, explanation, and prediction and not as an example of a mathematical structure" (Mandelbrot, 3).

Whatever its intrinsic aesthetic value, what links mathematics to science is instrumentality, that is to say, the ability of mathematical representation to enable a direct, practical intervention in the material world. Through all the twists and turns of complexity, quantum mechanics, and relativity, modern science has never lost confidence in the instrumentality of a mathematical representation of the world, and mathematical simplicity remains for scientists the highest heuristic principle (Heisenberg 1949, 59). However, at the time Terada and the other forerunners of complexity were contemplating this, there was no hope of verification by experiment or practical mathematical solutions, hence their efforts, often at the cost of their own careers, retain a certain aesthetic quality.

Resolving the notion of creativity at work in neuroscientific definitions makes it possible to assign a definite sense to the intuition that there is something "poetic" or "literary" about Terada's work. It strives toward complexity, and the irreducibility of phenomenal experience. It also supports the argument that the more speculative branches of complexity such as affordance, the science of forms, stratology, etc. verge on a paradigm qualitatively different from the reductive, analytical model of causality dominant in a reductive science, in that they seek to grasp a causality that comes from the totality or context.

This suggests that the pursuit of these new sciences will call on a different scientific disposition than that to which we are accustomed. Alluding to a passage from a book on "the physics of snow" by Terada's pupil Nakaya Ukichiro (1900-1962), the organizers of the 1996 Katachi U Symmetry Symposium sound again the classical split over which the disciplinary division of knowledge was overlaid in a generation in Japan, and the hermeneutic dimensions of the problem-consciousness that results:

> The character for katachi (形) is [also] used in Japanese for "metaphysical" and "physical," i.e., (形而上) and (形而下) respectively, which originally appeared in the classical Chinese work *I-Ching* (易経). The katachi is associated with many fields of human activity. Katachi is rich in information. "A snowflake is a letter to us

from the sky." "A diamond is a letter to us from the depths." Both statements allude to the multifaceted aspects of the concept of katachi. Looking back at some of the landmarks in the history of science, we can see that Mendel was aware of the katachi of peas in his pioneering investigation into genetics, and Wegener's theory of continental drift is based on observation of katachi. Paleontologists and archeologists draw information from the katachi of fossils and other remains. Good detectives, as well, draw information from the katachi in the evidence at the scene of the a crime. (Ogawa 2000)

The Hungarian physicist Denes Nagy, a proponent of the notion of Symmetry Studies calls on the Science of Form to bridge a gap between science and the arts as two hemispheres of a split culture (cited in Ogawa 2000). If Japanese scientists plowing new fields like morphological science and stratology feel themselves well-positioned to do this, it is in part because they have an example in front of them, in the peculiar figure of Terada Torahiko, straddling the line between literature and science, unable to commit to either side.

4 Incomplete Perspective: Deconstruction as a Technical Issue in Karatani Kōjin and Edogawa Ranpo

Karatani Kōjin is one of the few contemporary critics in Japan to have been translated to any extent into English, with two book-length essay collections: *Origins of Modern Japanese Literature*, published by Duke University Press in 1993, and *Architecture as Metaphor*, published by MIT Press in 1995. In the Foreword to *Origins of Modern Japanese Literature*, Fredric Jameson describes the collection as "one of those infrequent moments in which a rare philosophical intelligence rises to the occasion of a full national and historical statement," and expresses the hope that the publication of Karatani's book will have a fundamental impact on literary criticism. Less hopefully, he repeatedly underlines the provincialism of a Western literary critical establishment that, despite claims to have undone the centrality of the western canon, can't seem to incorporate a Tagore, Lu Xun or Sōseki into its map of major modern writers.

The close spacing of the publication of these two works is misleading, in that *Origins of Modern Japanese Literature* is a translation, thirteen years after the fact, of a set of tightly spaced essays from the late 1970s after a period of study at Yale under Paul deMan, while *Architecture as Metaphor* assembles and collates material from at least four different works spread out over the decade following *Origins*. Hence, the former represents the work which established Karatani at the beginning of the 1980s as a major spokesperson along with Asada Akira for Japanese postmodernity, while the latter represents his struggle through the 1980s and early 1990s to retain the critical force of that project while trying to come to terms with its relentless commodification as intellectual fashion.[1] This has involved Karatani in the introduction of math and technical issues into literary criticism, which results in a different relation to literature than that produced by the hermeneutically constituted genre of literary theory. This chapter speculates on the methods that might result through a reading of Edogawa Ranpo's short mystery "Hell of Mirrors" (*Kagami jigoku*, 1926).

By technical issue, I mean a material practice concerned with increasing performance. "Hell of Mirrors" is interesting because it brings together a host of problems relating to the institution and technical maintenance of a new visual culture. What allows this concentration is first an astute historical presentation of visual technologies, and secondly, the mathematical way the problem is posed. So the story "Hell of Mirrors" provides a chance to take literally the questions of construction and perspective that relate chapter 6 of *Origins of Modern Japanese Literature* to what I read in *Architecture as Metaphor* as an effort to sound the problems of deconstruction down an analytical line.

A Method to the Madness

The central intellectual tension of Karatani's work, prefigured in parts of *Origins* but emerging clearly in the episodic structure of *Architecture as Metaphor*, is the effort to think the deconstructive problematic he encountered at Yale outside its institutionalization as a specifically literary methodology in North America. That is to say, he would like to rescue deconstruction from literature. This has involved his work in a curious passage through the mathematician Kurt Gödel, touchstone of essay after essay in *Architecture as Metaphor*. Gödel's Incompleteness Theorem is presented as a sort of punch-line to a failed Western drive to formalization encompassing discipline after discipline of Western thought since the late nineteenth century, in which the very effort to build a rational "architecture" of knowledge and arrest the process of becoming, finds itself falling repeatedly into

---

1. The chapters that comprise *Origins of Modern Japanese Literature* appeared in installments in the literary journals *Kikan Geijutsu* and *Gunzō* between 1978 and 1980, and were published in book form as *Nihon kindai bungaku no kigen* (Kōdansha, 1980). The English language *Architecture as Metaphor* is drawn and reorganized from a number of essay and collections, including "Architecture as Metaphor (*Inyu toshite no kenchiku*, 1981), *Introspection and Retrospection* (1985), and the *Tankyu* series (Researches I, II, and III, 1985-1995).

self-reference and paradox. The first fifty pages alone introduce Kierkegaard, Alfred North Whitehead, the Pre-Socratics, Euclid, Merleau-Ponty, Binswanger, Dilthey, Galileo, Heidegger, Poe, Valéry, Baudelaire, Christopher Alexander, the Bourbaki group, Jakobsen, Mallarmé, Lao-Tze, Deleuze, deMan, John Locke, Yeats, Escher, Hilbert, Poincaré, Cantor and Ramsey. The implication is that the fields represented by these thinkers begin to "converge" in the late nineteenth century over the question of indeterminacy. Finding that the problematic of an indeterminacy looming at the edges of rational thought was well-known to the pre-Socratics and never lost to the Eastern tradition, *Architecture as Metaphor* goes back in history to investigate thinker after thinker, from the pre-Socratics and Hindu mathematicians to the logical positivists and vulgar Marxists, in terms of whether they confront the undecidability of the formal, self-referential system, a move that is read as formally identical to deconstruction, or whether they attempt to ground it, and turn it into an architecture.

Though *Origins of Japanese Literature* had an interdisciplinary character, and included chapters on landscape painting, interiority, sickness and hygiene, linear perspective, confession and the child, these were clearly understood as discursive constructs brought about through literature, and the writers discussed, from Mori Ogai and Sōseki, to Kunikida Doppo, Uchimura Kanzō, Tanizaki and Akutagawa Ryūnosuke, fit comfortably into a standard narrative of Japanese literature. In *Architecture as Metaphor*, though, the organizing concepts of "language, number and money" are discussed in terms of linguistics, mathematics, and economics, rather than their thematization or instantiation in literature, and interpolated with discussion of philosophy, architecture, city-planning, and poetry almost as an afterthought. This type of interdisciplinarity, where the other disciplines are present, has made his work difficult to classify in North America, where, despite his own identity as a literary critic, and despite the gesture to metaphor in the title and clear affinity to the project of deconstruction, *Architecture as Metaphor* does not register in the United States as literary theory, and is filed under architecture in bookstores and libraries.

The "Critical Space" style of argumentation as well,[2] which tends to use key concepts in the manner of stepping stones to leap between widely separated specialties and fields, has left Karatani open to the charge of intellectual carelessness. A style which can find for example the reflexivity of the circuit of capital in Marx to be identical to the antinomies of Kant as he does in the work on *Transcritique* from the late 1990s, or to equate the handling of oppositions like inside and outside, one and many, and subject and object by Mahayana thinkers 1500 years ago with the strategies of deconstruction in the late twentieth century, in which everything is an example of everything else, has to proceed to a degree by ignoring differences in context and theoretical detail. Jameson's own interdisciplinarity has been described as one in whose characteristic gesture "all theoretical differences evaporate into the dialectical sublime." The question raised of whether in such a case "we are dealing with an ability to synthesize derived from an intellectual power of a Lukácsian order, or merely a tendency to conflation," (Young 1990, 93) probably needs to be raised about the dizzying intellectual sweep of Karatani's own work, and is a project worth pursuing point by point with regard to *Architecture as Metaphor*.

Questions of execution aside, though, there is precedent for this stance in Japanese intellectual history, and a method to the interdisciplinary madness. Unlike Jameson, Karatani's tendency to equate thinkers widely separated in space and time does not reflect a Hegelian faith in their subsumption in the last instance, but rather the intuition that they are dealing with the same problems from the beginning. This position is not original to Karatani, but a well-established position visible in Japanese reactions to Western thought dating from the Meiji Period. According to a survey of Japanese philosophy by Miyake Yūjirō in 1909, the introduction by the American Ernest Fenollosa of the latest in German Idealism at the Imperial University in the 1880s was "not regarded with much surprise" by Japanese students schooled on the intricacies of Chinese and Hindu thought,[3] and the widely reported reaction was that they had seen this before:

---

2. *Hihyō Kukan* or "critical space" is the quarterly journal edited by Karatani and Asada Akira. The style of argumentation, where for example Heidegger, Nishida, Lacan and Zizek are dispensed with in the same sentence as reworking Kant's subject of transcendental apperception in their own terminology (*Hihyō Kukan* II-18, 25), is pervasive, particularly in the round-table discussions that open each issue, and has followed thinkers introduced there, such as Azuma Hiroki out into the wider journalistic world.

3. Fenollosa came from Harvard to lecture in philosophy at the Imperial Uni-

If the works of such men as Schopenhauer and Hegel are to be taken as the standard of pure German philosophy, we need not go to Germany alone, for there are many works in Hindu philosophy which treat of similar subjects. Though Hindu philosophy falls far behind German in minuteness of analysis, the former excels in many points of organization. . . . It is most probably that by scrutinizing the bulky literature of Hindu philosophy the greater part of pure German philosophy may be obtained. The difference between the Oriental philosophy based on the Hindu, and the Occidental based on the German, lies in the amount of scientific experiment. Eliminate experimental investigation and there is not much difference left between the Oriental and the Occidental systems. (Okuma 1909, 238)

Though one is used now to arguing the incommensurability of different cultural traditions, it is important to make sense of this reaction by non-Western intellectuals in which "there is not much difference" in the epistemology of contemporary Western thinkers and the Indian texts from a millennium or more before which they studied in their youth.[4] It is possible to argue that at a certain core philosophical level thinkers in even profoundly different traditions are talking about the same thing, i.e., manipulating a relatively limited number of concepts relating to time, space, the problem of error, the relation of the organism to environment, and to other organisms, the stability of meaning systems, etc., under different terminology. That for example the problem of representation comes up in every literate tradition, that similar skeptical strategies appear in ancient Greece and in India, that Nagarjuna and Parmenides pursue the problem of the one and the many to the same abyss, that when Augustine in the *Confessions,* and Dōgen in the *Shōbōgenzō* question the possibility of knowing past and future except as memory and expectation in the present they are talking about the same problem of time. This is how I understand the sense of recognition reported by Japanese students in the face of what was undoubtedly thought to be the pinnacle of

---

versity in 1880. His lectures on the history of philosophy from Descartes, and seminar on Hegel's *Logic* were remembered by students as "lacking in subtlety and exactness," and were said to give an impression "akin to that produced by scratching one's feet outside one's shoes" (Miyake 1909, 231).

4. It is worth noting again that the locus of identity of Meiji intellectuals when faced with the universalizing claims of Western thought is in Chinese and Buddhist

Western thought. The problem of indeterminacy posed by deconstruction is similarly fundamental, and the effort to find this problem
in a proliferating variety of fields is in direct continuity with the
philosophical stance of the Meiji Period.[5]

---

thought. The idea of an appeal to a particular Japanese identity does not become
prominent until the early twentieth century with the appearance in English of the
work of Okakura Tenshin, a student of Fenellosa's. See Anderson 1995 for a detailed
discussion of Fenellosa's role.

5. Though I have not seen it specified in print, the argument vis-a-vis deconstruction would presumably involve tracing the refusal of the deconstructive strategy
in the West back to Plato's commentaries on Parmenides, and juxtaposing this to its
fearless elaboration in Buddhist thinkers from Nagarjuna. The following dialogue
occurs in Plato's *Parmenides*, in which an older Parmenides steps in to admonish the
young Socrates on an ill-considered attack on Zeno. Socrates: "[T]here is nothing
extraordinary, Zeno, in showing that the things which only partake of likeness and
unlikeness experience both. . . . If however, as I just now suggested, someone were to
abstract simple notions of like, unlike, one, many, rest, motion, and similar ideas, and
then to show that these admit of admixture and separation in themselves, I should be
very much astonished" ("Parmenides," in *Plato I*, GB v. 7, 487). That is what Deconstruction sets out to do, to belatedly astonish Socrates. Parmenides begins his attack,
"I admire the bent of your mind, Socrates, but tell me now, was this your own distinction
between ideas in themselves and things which partake of them? . . . Let me say that
as yet you understand only a small part of the difficulty which is involved if you
make of each thing a single idea, parting it off from [other] things" (488). Parmenides
then, using a strategy of tracing out a series of oppositions until they dissolve into a
one-ness that strongly recalls the strategies of Hindu epistemology, twists the young
Socrates into a pretzel by showing him the difficulties he gets into in trying to affirm
that "ideas really are and we determine each one to be an absolute unity" (490).
"These Socrates, are a few, and only a few of the difficulties in which we are involved."
In this way Parmenides leads Socrates to the brink of the problems of deconstruction
and a putatively postmodern indeterminacy. However, in a conclusion in which he
reveals himself to be the mouthpiece of Plato, Parmenides backs away from the precipice:
"And yet Socrates, if a man, fixing his attention on these and the like difficulties, does
away with ideas of things and will not admit that every individual thing has its own
determinate idea which is always one and the same, he will have nothing on which
his mind can rest; and so he will utterly destroy the power of reasoning. . . . But then
what is to become of philosophy?" (490) The idea that there is no ultimate guarantee
for meaning, determinate ground for truth, no place where the mind can rest has been
unacceptable to the western philosophical tradition, and one can read the entire history
of western thought as a deferral of that problem, posed by Parmenides and the pre-
Socratics. However, the idea that there is no place on which the mind can rest did not
throw the Buddhist tradition into a panic. Rather the tendency for rational thought to
end up at this precipice was accepted as a key to the cessation of attachment to this
impermanent world. By the time of Nagarjuna, Buddhist thought had developed a
series of strategies by which such seemingly determinate oppositions as inside/outside,
self/ other, being/non-being and one/many are traced out in the most rigorous fashion
until they dissolve into paradox. That is why Buddhist thought often seems very

If there is good precedent for this in Japanese thought, there is also a method to Karatani's choice of fields that reflects its status as an intervention into literary criticism. Proceeding from a sense of recognition vis-a-vis the problem of deconstruction he encountered at Yale in the 1970s, Karatani situates Derrida of this period in a set of problematics peculiar to French discourse, and expresses dissatisfaction with the easy assent this commanded in certain precincts of the academy:

> The metaphor 'postal' itself begins with Lacan's discussion of Poe's "The Purloined Letter." It is a problem interior to the language games of the Lacan sphere of influence, and has next to no meaning for people who do not share this. . . . What Derrida is doing too is clearly within the Lacanian sphere, and is probably interesting in the interior of that discourse, but the question remains whether it is of interest to anyone else. In other words, will it be persuasive to people who are not convinced to begin with? Freud himself pointed this out as a limitation of psychoanalysis, i.e., what is psychoanalysis to do in the face of a person who will not enter into the transference? This applies just as well to the Lacan faction and to deconstruction. And when I speak of the "other" in *Tankyū I*, I am thinking of the other who is uninterested. . . . How does one persuade the person who is unconvinced? (Karatani, Asada, and Azuma 1998, 20)[6]

Karatani's immediate answer is "you just have to write clearly." But the substantive strategy developed gradually, seemingly by trial and error over the course of the 1980s, involves two parts. The first was to sound the strategies of deconstruction down the analytical line of philosophy. The second was to take the term "construction" from deconstruction literally, as in architecture or city-planning, in

---

contemporary. It has long embraced the kind of formal indeterminacy which most Western disciplines tried to defer until the nineteenth century. This at least, is how I imagine the argument would proceed.

6. This recalls Zizek's characterization of the transference involved in understanding Marxism: "You are not a Communist because you understand Marx, you understand Marx because you are a Communist," a mechanism he finds shared with Christian theology (Zizek 1989, 41, echoing Anselm). In a recent lecture at the University of Florida at the invitation of the English Department, Derrida seemed ironically aware of his status, referrring to himself as a preacher, and asking the audience to accept certain articles of faith prior to his discussion of ethical issues such as lying and the death penalty. Articles of faith are difficult to communicate outside the community of believers. ("The Future of the Profession, or the Unconditional University," Gainesville, FL, 12 April 2001)

order to introduce the question of technical practice into its formalized world.

The flirtation with analytical philosophy is hinted at in the appearance of Whitehead, Cantor, Ramsey, Russell and Hilbert in the first part of *Architecture as Metaphor*. However, the sense that these figures don't fit with such canonical theory points of reference as Marx, Saussure, Heidegger, Levi-Strauss, Lacan, Althusser and Derrida is not a function of their status or unfamiliarity, they are of course first rank figures in modern thought. It rather reflects a division of labor in the English-speaking world, by now well-remarked in standard histories of philosophy, wherein the "interest in the examination of purely theoretical minutiae and close ties with systems of formal logic" that characterize analytical philosophy are the purview of academic departments of philosophy, while the "practical commitment to the basic realities of human experience, work, power, love, and death" that characterize continental philosophy have become primarily the purview of departments of literature (Kenny 1994, 364). As introduced in the previous chapter, these two traditions can be distinguished historically in terms of lineage, "tracing each back, through G. W. F. Hegel and Gottlob Frege respectively, to their initial point of divergence in the nineteenth-century inheritance of the Kantian legacy." The general mechanism of understanding in hermeneutics involves relating parts to a whole:

> [W]e understand elements by correlating them within a larger context, which in turn must be reassessed, thereby provoking a reconsideration of its elements, and so forth. (Ray 1984, 54)

Opposed to this is an analytical notion that secure knowledge proceeds in an ordered progression from simple and self-sufficient parts to larger composite units. This leads to the interesting distinction by Conant that the Anglo-American or analytical tradition prefers to conceive of the world as a series of problems (or even puzzles), each in search of a 'solution,' while the continental, or hermeneutic tradition conceives of the world as a series of texts, each in search of a 'reading' (Conant 1991, 617).

Beyond making clear the provenance of Derrida's controversial "there is nothing outside the text," the significance of this point is that it makes immediate sense of the fact that the reception of Derrida and other French intellectuals has been mainly in departments of literature rather than philosophy in North America. The central intellectual tension in Karatani's work can be understood as the effort to sound the problematic of deconstruction developed in the continental tradition whose elaboration is excessively congenial to literary criticism, down the other, analytical line, where it still might surprise. The return by Karatani to an extended meditation on Kant in 1990s as *transcritique*, then, makes sense as part of an evolving project, in that Kant is the point of divergence between these two traditions.

In an essay called "A Map of Crises," by which the English version of *Architecture as Metaphor* is introduced, the architect Arata Isozaki situates Karatani within this tension between a deconstructive problematic and the idea of identity at some core philosophical level, in a way that recalls this study's concern with disciplinary boundaries:

> He is totally indifferent to the territorialities of today's scholastic subjects, which, though categorized arbitrarily, nevertheless have ended up constructing their own ivory towers and forming immutable and untrespassable boundaries; his writing traverses these boundaries as if they never existed. Unlike many multidisciplinary generalists who travel the horizontal strata, he transgresses categories by using his questioning to dig vertically through each domain while at the same time remaining within it. His procedure is consistent to the point of being violent; it is like a practice of pure radicalism. I cannot help but believe that he has a faith that some sort of original exists deep down.

I would argue that it is the inclination toward an analytic approach in a hermeneutic discipline, rather than the brute number of disciplinary boundaries crossed, which gives rise to the sense of heterogeneity in Karatani's work which Jameson and Arata recognize. Whatever criticisms can be made of Karatani's work from the perspective of individual disciplines, the call to think deconstruction outside its institution as a literary critical technique is original, finds good precedent in Japanese intellectual history, and is a part of Karatani's project that is still potentially subversive in a North American intellectual context.

Deconstruction as a Technical Issue

One of the first moves Karatani made in extricating deconstruction from its institutionalization as a hermeneutic strategy in North American departments of literature was to use linear perspective to figure the problem of *construction* in a technical, mathematical sense. This occurs in the chapter 6 of *Origins of Modern Japanese Literature*, and marks this chapter as a transitional point to the decade-long project of *Architecture as Metaphor*. I would like to outline what I understand the stakes to be in chapter 6 of *Origins*, and then turn to a discussion of the same question of perspective in Edogawa's "Hell of Mirrors."

As is well-known, *Origins of Modern Japanese Literature* is composed of six discrete essays each of which purports to show that a variety of categories taken to be natural in modern literature, were actually instituted over a short period of time in the Meiji 20s and 30s (roughly 1887-1907), through the introduction of a variety of material practices. Though not explicitly stated, it would seem that Karatani's gesture, insofar as he is concerned to show that categories such as landscape, psychological interiority, sickness, children and depth, are not "just out there," but are the product of historically conditioned *forms* the modern mind brings to bear on the manifold of experience, operates throughout in the mode of Kant's Copernican revolution in the First Critique. Except that he is arguing the a priori forms of cognition are not universal, but historical. And so, Natsume Sōseki's ironic comments about his failure to appreciate the universal appeal of Shakespeare despite years of effort reveal that "his real concern was to point out that the universality claimed by English literature was not a priori, but historical" (12). Kant's Copernican gesture in the First Critique was, in response to the notion that our thoughts, perceptions and concepts take the form they do because that's how objects are, i.e., that the forms of our cognition are imposed from the objective world, to reverse the directionality and say, what would happen if we assume that the form of objects is imposed by the perceiving mind? What begins Karatani's reflection on the idea that the seemingly self-evident category of landscape might conceal an inversion is that the Meiji writer Kunikida Doppo (1871-1906) discovers his "unforgettable people" (*wasureenu hitobito*, i.e., people in landscape) in the quiet hours of his study.

> It is only within the 'inner man,' who appears to be indifferent to his external surroundings, that landscape is discovered. It is not perceived by those who look 'outside.' (25)

This can be compared to Kant, in regard to the basic problem of error, of whether we can be sure the world really takes the form in which it appears: "[Y]ou have no further to go to discover the answers, for they lie entirely within" (Preface to the 2nd Ed. First Critique). Karatani takes a laundry list of bourgeois verities assumed to simply exist waiting for description: in successive chapters landscape, interiority, sickness, children, and space, and says, what if there are no such things "out there," but that these are *forms* the modern subject imposes on the world? That is to say, he seeks the form of social categories not in a pre-existing reality, but in the discursive space in which they acquire sense.

What intervenes in the mid-Meiji period to effect this overturning of sensibilities is the wholesale importation of a host of material, technical practices such as compulsory schooling and universal conscription with their associated communal practices, techniques of hygiene, everyday practices of work, communication, transportation, etc., linked to individual cognition through representation in art and literature. The discrete essays of *Origins of Modern Japanese Literature* go on to persuasively demonstrate that over a relatively short period of time in the second half of the Meiji period, people looked around and suddenly started to find children everywhere, suddenly found themselves in landscapes, writing literature, suddenly felt the need to confess, and that once the epistemological constellation (*ninshikirontekina haichi*) that allowed the perception of these natural objects was in place, it was only transitional figures like Sōseki and Tsubouchi Shōyō, who could perceive their historicity. In certain modern writers who straddle exactly the institution of modern literature as a system in the third decade of the Meiji period, Karatani sees an inability to place Western forms "in perspective," that is, as the telos of a historical progress, or in terms of a universal value. Instead, Shōyō and Sōseki juxtapose Western and Japanese (and Chinese) literature in a decentered structure and analyze them formally. That is why they suddenly seem fresh again today.

> Western writers have only achieved this perspective after entering a period when the West has been decentered vis-a-vis the rest of the world. (Karatani 1993, 148)

That is to say, because the system of modern literature was instituted within a generation in the third decade of the Meiji period, it exists as a lived contradiction within individual writers (which places them in a position of alterity to modernity that is formally like postmodernism). For the new generation of writers and intellectuals, though, they rather appeared like self-evident and natural categories.

The argument is repeated through each chapter, as seemingly self-evident parts of modern life are found to lie internal to literature as a configuration. Here it seems that the enigmatic term "epistemological constellation" is to be understood as an arrangement (*haichi*) of a priori universals. That is to say, it is the *arrangement* that is

historical, put into place through material practices from standardization of the written language to primary school reforms, which like
the tumblers of a lock falling into place, suddenly open in the Meiji
period, and children appear. Of course there were humans between
the ages of 2 and 18 in the Tokugawa period, people got sick and
died, walked on the beach or through forests, and reflected on their
experience, but there was no unified discursive space in the fractured,
heterogeneous configurations of the Tokugawa period in which these
experiences could appear as a coherent category.

In demonstrating how an entire *arrangement* of universals fell
into place in the Meiji 20s, so that suddenly everyone found themselves
under it, a large burden accrues to Karatani to show the systematic
nature of its institution. It becomes clear in retrospect from the sixth
chapter that the schema he uses to mediate this shift over a whole
range of experience, spaces, practices and phenomena is another technical practice: the production of the illusion of depth through linear
perspective. The particular form he sees instituted over the Meiji 20s
across a wide swath of social existence is the gathering, organization
and reduction of heterogeneous instances to a single transcendental
point: the vanishing point, figured as the Christian God.

Hence Uchimura Kanzō, the famous Christian thinker finds the
pantheon of contradictory gods to which he prayed as a child ordered
at a stroke by his introduction to Christianity. Sickness is referred to
the "theological cause" of the germ, and the heterogeneity and play
of surfaces in Tokugawa art and literature are referred to illusions of
three-dimensional space and the depth of psychological interiority.
The argument acquires its systematic nature because in dealing with
composite social categories, as opposed for example to space and
time, linear perspective can only exist as metaphor, as the abstraction
of a particular representational logic. That is to say, the argument
rests on extrapolating a logic drawn from the problem of space as an
a priori category of cognition to social or discursive space.

Chapter 6, entitled "On the Power to Construct," begins:

> When we read so-called premodern literature, we often have the
> feeling that it is lacking in depth. Since people of the Edo period
> lived under routine exposure to a wide variety of terrors, ep-
> idemics, and famines, however, it is unlikely that they were not
> feeling things deeply. If, in spite of this, we say there is no depth
> to their literature, what can this mean? . . .
>
> The question is perhaps more comprehensible if we transpose
> it to the area of painting. Japanese premodern painting also appears

to lack something like depth; to wit, it lacks linear perspective. But the linear perspective we now see as natural because we have long been accustomed to it was originally not a natural thing. . . . The depth that we see is based on a system of drafting, consolidated over centuries of effort, largely in the area of mathematics rather than art. (Karatani 1993, 136)

Karatani's reliance on Erwin Panofsky's seminal work on perspective becomes clear with this passage. In "Perspective as Symbolic Form" (1924), Panofsky finds in perspective a schema that links experience in general with cognitive experience and technical practices. Because the elaboration of perspective in art is dependent on technical practices, this provides a way to ground and mark the historical transformation of sensibilities in a given society. Linear perspective, then, the product of a centuries long, cumulative effort to banish inconsistencies and produce a rational, that is abstract, homogeneous and infinite space, is the unconscious expression of the modern sensibility.

It is clear from this that in some sense Karatani's work in the 1980s is an extended commentary on Panofsky. What Karatani seeks to do in the unique "laboratory" of Meiji Japan is raise the question: What would happen if these technical practices were imported wholesale into a society that lacks the "will to form" that Panofsky posits behind the Western expression of perspective? That is to say, the centuries long consolidation of technical practices that produced the sensibility of the modern West is imported into Meiji Japan at a stroke. However the Western "will to form" is not. Hence in *Origins of Modern Japanese Literature*, he finds in Japan after the transition of the Meiji 30s a society with a fully modern sensibility, but without the desire to "tie the strands together" and banish inconsistencies that characterizes the West. Hence transitional figures like Sōseki, Shōyō and Mori Ogai "slip back" after impressive displays of modern consciousness. And new genres like the I-Novel putatively centered on the new modern subject end up devolving into a centerless interrelationship of fragments. The subsequent decade of work gathered in *Architecture as Metaphor*, then, casts this whole drama as internal to the West, where the central problematic of the twentieth century is found in the reassertion of a "ceaseless becoming" in the face of a "will to Architecture" that has dominated the West since Plato. The unique historical circumstances of Japan implicitly place it in a position where it has already achieved this postmodern perspective.

The transition to chapter 6 of *Origins of Modern Japanese Literature*, then, represents a point at which the enabling hypothesis of an evolving project suddenly comes together, and a sense of a shift is almost palpable. Because the first five chapters take as their object complex higher-order categories like interiority, sickness, landscape, the child, once one gets past the paradoxical formulations, there is something obvious about the assertion that these are historical, what else could they be? The operation by which a discursive system or "semiotic constellation" would snap into place, and suddenly children *appear* over a prior heterogeneity seems a level removed from the directness by which space or bodies are perceived, i.e., a level removed from cognition. Chapter 6 begins with the same move, to say that the sense of "depth" and three-dimensionality we get from modern literature and art is not to be construed as an attribute of reality "out there," it is rather the product of a particular configuration, construction by the technique of linear perspective. But with depth in the literal sense of a spatial dimension, the argument about historicity reaches down to one of the a priori forms itself, and the Kantian strategy swallows its tail. In Panofsky's terms, an argument about categories of experience takes as its object a category generative of experience. Hence, a wavering begins between literal and figural senses of space, prefiguring the wavering between figural and literal senses of "architecture" that is the central conceit of *Architecture as Metaphor*, and it is symptomatic that this chapter only splits into two.

It is a commonplace now in literary theory to challenge the hierarchy between figural and literal implicit in assertions of the form "X in the literal sense," whereas scientists and engineers tend to grant priority to certain domains, such as, for example, the physical world. So it is important to specify what I mean by saying the introduction of the problem of linear perspective in chapter 6 brings up space in its literal sense. I take space to mean the determination by our senses of the position of bodies. I take terms like perception, perspective, depth, etc. used in relation to physical space to be prior to metaphorical uses like "discursive space," "space of debate," or "psychological depth," in that the more familiar and clearly-delineated idea is used to perceive and investigate the less clearly delineated senses of space in the latter. Any metaphorical relation has this implied hierarchy between source and target domains, and these are open to challenge. Space in this sense can be argued to be prior, though, following Quine in *From Stimulus to Science* (1995), in that a schematized intuition of space is a necessary part of our picture of the world, for understanding the relations of objects, and the implied trajectory of bodies when we can't see them. This schematization of space, fragmented, inconsistent, and constantly shifting comes from the relation of organism to environment, is in some form necessary for survival, and gains some regularity across a group by practice in using spatial connectors like "above" and "below" (Quine 1995). Technologies like linear perspective are important not because they give an accurate reflection of subjective experience, but because they sharpen and rationalize this fragmented, sketchy, shifting sense of space we acquire through interaction with the environment. In this it is *instrumental*, and, as Meiji thinkers such as Miyake Yūjirō and Sōseki clearly recognized, these sorts of technical issues related to repeatability and performance are what separates modern science from other types of science.

Karatani's discussion of linear perspective is of interest because, while using it to challenge the idea that the sense of "depth" one feels in regard to the well-drawn characters of modern Japanese literature is in any way natural, he never loses sight of the status of linear perspective as a technical, mathematical issue. This allows him ultimately to resist the temptation to dismiss perspective as "arbitrary" in its relation to the subjective sense of space, thereby sharpening the point that it is arbitrary in aesthetic matters.

The aspect of Panofsky's essay that has attracted most attention outside the field of art history is its ambition to relativize the claim that the illusion of depth in linear perspectives is natural. "[T]he promise of undermining the claims to legitimacy or naturalness of linear perspective . . . has always been the basis of the perspective essay's celebrity" (Wood, in Panofsky 1997, 22). However, in the end, need for Panofsky to introduce the idea of an "internal visual image" to systematize the relations between different unified systems of perspective, for example Greek, Medieval and modern perspective, "brings down one of the most sensational ambitions of the essay."

> Whether or not perspective is in fact an arbitrary convention . . . Panofsky in any case fails to fulfill his own promise; indeed he rather quickly backs off from extreme relativism. (Wood, 22)[7]

---

7. In first posing, then backing off from the claim that linear perspective is arbitrary, Panofsky sets a pattern that persists to this day in discussion of realism in the humanities. See for example Heath (1983), 86-89, who, in a reply to Noel Carroll presents one of the most lucid discussions of the difference between natural perspective

According to Panofsky, perspective's interest, and the reason it makes a promising case study for grasping the modern sensibility, is not because it describes the world correctly, or the way the world appears in our subjective impression, but because it describes it according to a rational and repeatable procedure. "That is precisely the enormous advantage of the modern method, precisely why it was so passionately pursued" (40). In *Art and Illusion*, Ernst Gombrich pursues a similar strategy, arguing against the counterintuitive position that linear perspective is in an arbitrary relation to perception. For Gombrich, perspective proceeds from the fact that light can't go around corners. "This is all perspective can and does claim" (254). That perspective is different from other representational strategies not because it represents the way things are in the world, but because it captures the way light gathers in the retina.

> Here perhaps are the inarticulate roots of he idea that perspective is merely a convention and does not represent the world as it looks. Perhaps also . . . the wish for a stick with which to beat the Philistine who wants to have his picture 'correct.' (Gombrich 1960, 254)

Both points need to be acknowledged in dealing with the question of 'realism' in glass-based representational technologies. First, that there is a serious mathematical issue of a rational and repeatable procedure, and second, that this set of relations can be mapped onto the physical world at the point of the convergence of light in the vertebrate retina. Karatani acknowledges these points in a way that marks his as a mathematically informed treatment; acknowledging the technical feat of linear perspective, but finding it arbitrary in aesthetic matters.

> Two points emerge from the foregoing discussion: First, that it is for want of a configuration producing such a sense that premodern literature lacks "depth," and secondly, that the presence or absence of such a configuration can in no way be used to determine literary merit. (137)

There are many other forms of perspective that are rational and repeatable, as for example the open perspective of *ukiyo-e* prints, or

---

(i.e., the laws by which images are generated on the retina) and linear perspective available, and in the process backs away point by point from the more radical claims of the arbitrariness of their relation.

the orthogonal perspective of engineering graphics, however they do not capture this point. There are many simplifications and abstractions involved that distance linear perspective from perception in the curved surfaces of binocular, moving vision. But the particular technical trick and its basis in optics stand. This is why the idea of linear perspective as a schema structuring the modern sensibility makes a promising case study for elaborating a deconstructive project. As a rational and repeatable procedure, perspective invites itself to be pursued to the point of self-destruction.

Edogawa Ranpo and the Taisho Obsession with Visuality

Karatani's argument dates the institution of a modern, perspectival sensibility to the late Meiji period, say from 1890-1910, hence it makes a kind of sense that one finds in writers of the Taisho period (1912-1926) a wholesale importation of image-making technologies into their stories. The mirror, the lens, and the movie camera, like the window of perspective have an abstract existence as metaphors for modern knowledge and subjectivity, however, they also existed as a procession of glass-based technologies which enabled specific material practices, and these were flourishing during the Taisho period. Stereoscopes, magic mirrors, panorama, telescopes and photography appear repeatedly as a thematic concern in the stories of writers like Uno Koji, Sato Haruo, Murō Saisei, Izumi Kyōka, Tanizaki Junichiro, and in the author's private lives as well. This might seem a set of dilettantish tastes and aesthetic affectations, assimilable to the solitary imaginative world of the elite male writer. However, if glass-based technologies provide a site at which, as Jonathan Crary suggests, perspective as a discursive formation intersects with material practices, it may also represent a grasping by these writers to trouble from within what Karatani characterizes as the modern perspectival regime of representation, to bring the process of seeing into this homogeneous visual field (Crary 1988).

The way the technical problem of linear perspective becomes related to a deconstructive logic comes out with particular sharpness in a short story by the mystery writer Edogawa Ranpo (1894-1965) from 1926 called "Hell of Mirrors" (*Kagami Jigoku*, 1926). Edogawa takes the problem of perspective in much the same context and form Karatani does (as an instantiation of a Western will to form), and pushes its protocols to the moment of self-reference. In pushing the image technologies of his day to the question of complete perspective, Edogawa pinpoints the link between the historicizing project of *Origins of Modern Japanese Literature* and the logical clash of *Architecture as Metaphor* perhaps more precisely even than Karatani.

"Hell of Mirrors" tells the story of Kan Tanuma,[8] a middle-school graduate of independent means who evinces from an early age an obsession with optical and reflecting devices.

> Some might have called him just eccentric, but I am convinced he was a lunatic. At any rate, he had one mania—a craze for anything capable of reflecting an image, as well as for all types of lenses. Even as a boy the only toys he would play with were magic lanterns, telescopes, magnifying glasses, kaleidoscopes, prisms, and the like. (Edogawa 1956, 109; 1991, 239)[9]

Addicted to the production of images, Tanuma pursues his obsession through a series of increasingly invasive and sophisticated installations and displays, and descends slowly into a hell of voyeurism, narcissism, and fevered hallucinations. In the end, surrounded by servants and a single witness, he encloses himself in his final creation, a spherical mirrored enclosure and emerges some hours later, completely mad.

"Hell of Mirrors" is interesting because it brings together, in one compact package, a virtual catalogue of the problems relating to the institution and maintenance of visual culture: technologies for seeing, architectures of vision, voyeurism, narcissism, performance, the erotic, ocular regimes, optics, blindness, displays, exhibitions, installations, cartographies, invisibility, blindspots, vanishing points, the chemical, the alchemical, perception, projection, disclosure and illusion.[10] What allows this concentration is first, an astute historical presentation, and secondly, the technical way in which the problems are posed.

The problem, or "mystery" of the story, is of course what the protagonist saw inside that final project, the spherical mirror, that made him go mad.

---

8. He is referred to only as "my friend K" in the Japanese.

9. I have used the rather loose translation available in *Japanese Tales of Mystery and Imagination* (Tokyo, Tuttle Co., 1956), 108-122, modified where necessary. The original can be found in *Chikuma Nihon Bungaku Zenshu — Edogawa Ranpo* (1991), 238-266. Citations will give the page numbers in the English edition followed by page numbers in the Japanese, in the form: (eee; NBZ: jjj). Tuttle has reissued the 1956 translation with the same pagination.

10. This is actually a set of suggested topics for submissions to a new *Journal of Visual Culture*, (e-mail message to KineJapan mailing list, dated 21 Mar 2001, 01:36:57) however one could easily go through Edogawa's story and point to each topic.

I continued to stand alone in the laboratory, my eyes fixed on the glass fragments scattered about the room, desperately trying to solve the mystery of what had happened. For a long while I stood thus, wrestling with the conundrum. . . .
Why did he have to go crazy? No, more than that, what could he have seen inside that glass globe? Exactly what did he see? (121-122; NBZ: 264)

The Gothic dimensions of the story are obvious. Obsession with mirrors is a standard figure in late-Victorian Gothic literature from Paget's *Amour Dure*, to Walter Pater and William Butler Yeats. Edoga-wa combines here the "fundamentally divided Paterian impression-seeker" with the concern for a perfect reflection Yeats recalled in a friend:

He thought to spend his life, insofar as it was an artistic life, in making the silver mirror without speck. (Yeats, Memoirs, quoted in Block 1993, 36)

Scottish essayist Thomas Carlyle writes in *Past and Present* that the Hells of people differ notably, locating Hell for the English in bankruptcy, however "[t]he word Hell . . . generally signifies the Infinite Terror, the thing a man *is* infinitely afraid of, and shudders and shrinks from, struggling with his whole soul to escape from it" (quoted in Weiss 1986, 13). In "Hell of Mirrors," Edogawa has in mind the latter, the Infinite Terror.

Perhaps one can reduce the final madness to a question of the deleterious effects of the mercury evidently used in processing the mirrors. In the final sequence, the protagonist commissions a select staff of technicians and workmen to produce the glass sphere. Conditions of factory work emerge in a detail:

Of course the technicians wouldn't have known the purpose, but they were ordered to paint the exterior of the globe with quicksilver, in order to make the interior a one-unit mirror. (121; NBZ: 263)

The colloquial phrase "mad as a hatter" refers to effects produced among workers at the turn of the century by the use of mercuric nitrate in the manufacture of felt hats. These effects included hair and teeth loss, loss of memory and a general deterioration of the nervous system (Stwertka 1996, 182). And Lombroso, the turn of the century phrenologist introduced to Japan through Sōseki, lists it in

The Man of Genius as one of the points where genius and insanity meet in a degenerative process.

> A theory, which has for some years flourished in the psychiatric world, admits that a large proportion of mental and physical affectations are the result of degeneration, of the action, that is, of heredity in the children of the inebriate, the syphilitic, the insane, the consumptive, etc.; or of accidental causes, such as lesions of the head or the action of mercury, which profoundly change the tissues, perpetuate neuroses or other diseases in the patient, and, which is worse, aggravate them in their descendants, until the march of degeneration, constantly growing more rapid and fatal, is only stopped by complete idiocy or sterility. (Lombroso, 1895)[11]

However, one must observe that Newton repeatedly refers, in the *Optics,* to the use of quicksilver in his experiments, and a more historically astute and philosophically respectable context might be found in connecting this "infinite" terror to the question of perspective, suggested by the attention Edogawa gives to the historical procession of image-making technologies. The idea that linear perspective is the expression of a schema linking social, cognitive and psychological experience with technical practices argues that institutions have organized themselves and mobilized individuals to take their place in them through a model of vision that implies a clean split between the subject and object of vision, and a rational, that is abstract, infinite and homogeneous space secured by a transcendental gaze. This is the modern visual regime. In *Techniques of the Observer,* though, Jonathan Crary makes a distinction between a "geometrical optics" in the seventeenth and eighteenth centuries (i.e., a perspectival science of correctly producing objects) and a "physiological optics" beginning in the nineteenth century (a science that looked into vision itself, and how it produced objects), in which technologies drive the shift. The technologies of mirror, camera obscura, etc., produced a perspectival image perceived as separated from the viewing subject. However devices like the stereoscope which emerged in the nineteenth century, by producing through correct perspective the obvious illusion of a solid body, called attention to the process of vision, to the beholder's

---

11. Quoted in *Bulletin of Psychology and the Arts,* v. 1 (2), 30. Cf. Sōseki's striking comment in *Tower of London* that a prolonged stay in that city might well "turn my nerves into a state like heated glue."

share, to the subject's participation in the object. That is to say, vision became the object, rather than the subject, of optics.

If Karatani is correct in seeing linear perspective (i.e., the seventeenth and eighteenth-century geometric notion) as the paradigm instituted in modern Japanese literature in the late Meiji period, then the documented fascination of Taisho writers with visual devices (cameras, magic lanterns, stereoscopes, cinema) can be read as an attempt to effect that shift, to use image-making technologies produced by these institutions to destabilize their orders, create heterogeneous spaces, and act in ways that could not be anticipated. And because of the way the wholesale importation of the technical practices supporting the former regime was overlaid by the technical practices supporting the latter, early twentieth-century Japan becomes one of the densest sites for exploring this kind of encounter between technology and sensibilities. That is to say, behind the light Gothic intoxication of Edogawa's story may be an intensely serious trajectory.

The question, of course, is what he saw inside the spherical mirror. The key is offered early in the story in a curious reference to a "heredity" to the affliction Tanuma carries. This legacy of an affliction (*iden shita no wa*) is, as the context in nineteenth-century discourse on "genius" suggests, not a genetic inheritance, but an inheritance of acquired characteristics, in this case a grandfather or great grandfather's conversion to Christianity, as the narrator specifies in the second paragraph, which bequeathed to Tanuma a fund of strange images, supernal fetishes, images of Mary and the crucifixion, telescopes and spy glasses. In a pattern familiar from the discussion of Sōseki's *Tower of London* in chapter 2, however, the three passages which specify the origins of Tanuma's obsession in Christianity are omitted by the English translator as unnecessary detail.

Tanuma's procession begins then, with a polished metal mirror, also a family heirloom.

> One day, when I visited him in his room, he had out on the top of his desk an old *kiri* box, and I suppose he'd taken it out of there, but in his hand was a very old metal mirror, and he was turning it in the sunlight and producing shadows on the dark wall. (NBZ: 240)

Unfortunately, this episode is omitted from the English translation as well.[12] The polished metal mirror evokes a premodern visuality of

---

12. The episode takes over two pages in the Japanese text, glossed in the English

natural materials, the age of the gods, when the material of the mirror was part of the process of reflection, and one can think here of the water in which Narcissus finds himself reflected, which ripples at the touch. The juxtaposition of this tangible premodern visuality with the glass technologies from the heretical Christian tradition indicate a period in the past when such models of vision could exist side by side. Here is the point where the location of Tanuma's "affliction" in heredity becomes crucial, because what the turn to the Christian legacy introduces is the abstraction from materiality and obsession with the transcendental point that characterizes a linear perspectival system. The metal mirror tarnishes, it has surface imperfections, it dilutes the image with its own materiality. The transition from the polished metal mirror to precision, glass-based technologies indicates the desire for successively more perfect reproductions, to "make the silvered mirror without speck."

As Tanuma advances into higher education, he embarks on a succession of investigations that grow progressively more manic and obsessed, and gain a basis in science:

> He always had a craze for mirrors and lenses, but this turned over into a lens-mania that verged on disease when he got into middle-school, As you know, physics has a good deal of theory about lenses and mirrors, and when he began to take physics as an upperclassman he became completely preoccupied . . . (109; NBZ: 242)

The rest of the story, then, consists of a series of episodes in which Tanuma seeks to gratify his near narcotic need for images through increasingly elaborate experiments with mirrors and lenses. The first step occurs when he encounters a concave mirror in class, and lets out a shriek of joy at the magnified image of his face.

> Subsequently, his love for concave mirrors grew so intense that he was forever buying all sorts of paraphernalia—wire, cardboard, mirrors, and the like, from which he began constructing various devilish trick-boxes with the help of many books he had procured, all devoted to the art of scientific magic. (110; NBZ: 244)

Generously supported by his anonymous parents, Tanuma fore-goes advancement to the Higher School, and when both die in his

---

translation as "episodes concerning Tanuma's craze for mirrors and lenses in his boyhood are almost endless . . ." (Edogawa 109; NBZ 19, 240-242)

20th year, he uses the inheritance to build a research-level telescope in the garden, whose powerful resolution he immediately trains on the surrounding households.

Now completely free from any supervision, and with ample funds to satisfy his every whim, he began to grow more reckless than ever. (111; NBZ: 246-7)

Subsequent steps in Tanuma's odyssey include a special type of periscope, which he uses to get a look at his maidservants in his lab, a stereopticon, by which he projects a grotesquely magnified and inverted image of his living face on the wall, a small room lined with mirrors on all four sides and ceiling and floor in which he spends unseemly amounts of time with himself, and with a maidservant consort. Exhausting his fortune, "he kept laying in bigger and bigger stocks of mirrors of all shapes and descriptions—flat mirrors, concave, convex, waveform, prismatic—as well as miscellaneous specimens that cast completely distorted reflections" (114; NBZ: 254).

Finally, Tanuma reaches a point in his investigations where he could not press any further with commercially available materials, and so he hires on a staff of technicians and workmen to establish a glass-working plant in his garden. Through this workshop he produces wall-sized mirrors, and a gigantic kaleidoscope which can be entered and viewed from the inside. These large-scale installations, involving Tanuma dancing around in the laboratory among the shards and shapes of glass "like the flowers of an opium addict's dream" and wondering at the reflections and projections on the laboratory wall, surround the protagonist more and more completely with reflections of his own image, and bring the story on trajectory to denouement.

Nor was that the last of it—far from it. His fantastic creations multiplied rapidly, each on a larger scale than the previous one, pushing any human who saw them to the limit of their sanity. And shortly thereafter came the terrible, tragic climax. (115-116; NBZ: 257-258)

Though seemingly determined by his sexual awakening, and "*ero-guro*"[13] need for stimulation through voyeurism and display, the series conceals, as indicated by its basis in the study of optics, a systematic play with the problems of perspective. That is to say, there is a method

---

13. "Erotic-grotesque." Connotation is similar to "flappers," as a way of referring to the alarming addiction to stimulation in urban youth of 1920s Japan.

to Edogawa's madness. As a technical problem in art history, linear perspective involves a tension between the essentially mathematical problem of the expression of the concept of infinity in a Cartesian, that is homogeneous and rational, space, and the artistic requirement to represent images drawn from human experience. Hence, linear perspective has been plagued since its inception with two problems: discrepancies between the homogeneous space of linear perspective and what is given in perception, and distortions at the margins as the mathematical grids move out to infinity. In cataloguing the "rather bold abstractions" involved in its producing a rational and repeatable procedure, Panofsky writes,

> [I]t is not only the effect of perspectival construction, but indeed its intended purpose, to realize in the representation of space precisely that homogeneity and boundlessness foreign to the direct experience of that space. In a sense, it transforms psychophysiological space into mathematical space. . . . It forgets that we see not with a single fixed eye but with two constantly moving eyes, resulting in a spheroidal field of vision. It takes no account of the enormous difference between the psychologically conditioned 'visual image' through which the visible world is brought to our consciousness, and the mechanically conditioned 'retinal image' which paints itself upon our physical eye . . . [and] perspectival construction ignores the crucial circumstance that this retinal image—entirely apart from its subsequent psychological "interpretation," and even apart form the fact that the eyes move—is a projection not on a flat but on a concave surface. (Panofsky 1997, 31)

Edogawa addresses these abstractions almost point by point. Moving beyond the material imperfections of the polished metal mirror, one of the first things the protagonist begins to investigate is a concave mirror, that is to say a mirror which captures the concavity of the retina, and he is fascinated by what he sees. The stereopticon is a method for producing the illusion of depth by projection from two points, and addresses the fact that we see with two eyes and not a single eye. Gombrich glosses the particular optical insight rationalized through perspective as "the fact that the eye cannot see around corners." It is this optical point that is shared with mirrors, and the next step in Tanuma's systematic investigation of the possibilities of glass-based image technologies, is the periscope, which enables him to see around corners. The chaotic dancing, and immersion into room size mirror installations too, while seemingly introducing a corporeality, allow the protagonist to abstract himself from the tactile experience of space, surrounding the body with a boundless succession of images, "foreign to the direct experience of that space."

In bringing the reader step by step through a historical series of perspectival technologies, though, Edogawa soon runs out of the problem of the limitations of the perspectival image, and into the question at the center of the drive to the perfect copy, the God's eye view. As a technique of construction, linear perspective is a system that catches the image on a flat surface, a canvas, Alberti's "window of perspective," and this is one of its weaknesses, producing distortions at the edge familiar from photography. One of the major technical problems in the development of perspectival systems, a limit case of these marginal distortions, is the mapping of the earth on a two-dimensional surface, as in the Mercator projection where the map of the earth is projected onto a flat piece of paper, but at the outer edges one sees strange cuts to the paper, Greenland oddly distended, and the shortest distance between two points is no longer a straight line.

> Finally, with the first globe (1490) and the invention of the Mercator projection around the same period, yet a third dimension of cartography emerges, which at once involves what we would today call the nature of representational codes . . . and in particular the unresolvable . . . dilemma of the transfer of curved space to flat charts. (Jameson 1993, 90)

The problem expressed here is the problem of viewing the totality of the world from a single point, and it has vexed mathematicians and mapmakers from the beginning of the modern period. The protagonist's final move in capturing a perfect perspective is to enclose himself entirely in a spherical mirror.

> A long time ago Tanuma had ordered the chief engineer to construct this glass sphere. Its walls were half an inch thick and its diameter about four feet. In order to make the interior a one-unit mirror, Tanuma had the workmen and engineers paint the exterior of the globe with quicksilver, over which they pasted several layers of cotton cloth. The interior of the globe had been built in such a way that there were small cavities here and there as receptacles for electric bulbs which would not protrude. (121; NBZ: 263)

Tanuma enters the mirror one night, and emerges from the mirror hours later, completely mad. The question raised to the reader at the end, as the narrator ponders the shards of glass scattered over the floor of the laboratory, is of course what the protagonist *saw* while inside the mirror. This sort of rhetorical question is conventional within the mystery genre in which Edogawa writes, and the properly literary response would seem to be to allow it to provoke a moment's pleasant reverie.

Simply elaborating the logic of the procession, what the protagonist *does* when he enters that spherical mirror is solve at a stroke the problem of totality that had vexed centuries of mapmakers and technicians of perspective, and the reason that he goes mad is that in taking up a position where he can survey the totality from a single point, *he becomes God*. That is to say, what Edogawa has done very cleverly in putting the legacy of this sickness in his grandfather's Christianity is to place the path to this "fall" into Hell in the pursuit of a modern perspectival regime in Meiji. The *optical device* (Tanuma's string of obsessions) is being introduced into the *discursive regime* (Karatani's master schema of linear perspective), within which it deconstructs. It is the Taisho visual strategy par excellence, pressed to its conclusion.

One can look in this regard at scholastic attempts to grasp the infinity of God. God for the later scholastics is not the God of popular imagination, but the singularity of a mathematical system, demythologized as pure thought contemplating itself to eternity (Blumenberg 1987, 437-487). When Tanuma enters the spherical mirror, and surveys the totality from a stable point, he actually becomes God, and the reality from which he came is just a fold. From the limited perspective of the servants outside he emerges X hours later, but it has been infinity. What is a matter of minutes for the others, is for him a timeless domain where, as God, he sits contemplating himself, to eternity, and he emerges insane.

$$\frac{1}{0} = \infty = \text{pure many without one} = \text{chaos}$$

In *The Fold*, Gilles Deleuze relates these two reciprocal terms, ∞ and 0, with God:

And God, whose formula is $\frac{\infty}{1}$, has its reciprocal in the monad, $\frac{1}{\infty}$.

(Deleuze 1993, 49) For scholastic philosophy, as it gave way in the seventeenth century, zero and infinity were attributes of God. They were singularities in the system of calculus, and in the completeness and continuity of real numbers, entities outside the domain of human experience. Inside the sphere, the mundane experiential problems of the mirror image give way, and one experiences the full force of the concept of the infinite, bound to trail off into distortion in a two-dimensional image. This, that is to say, reflection on the mathematical problem of zero and infinity that linear perspective tries and fails to grasp, is the context for reading Edogawa.

The Problem of the Spherical Mirror

Given the conventions of the mystery genre, we have allowed the formulaic ending to provoke a moment's reverie, and come to some conclusion about what the protagonist *does*. But oddly, we reach a similar conclusion if the question is taken literally.

> What, then, I asked myself again, had he seen? It was surely something completely beyond the scope of human imagination. Assuredly, never before had anyone shut himself up within the confines of a mirror-lined sphere. Even a trained physicist could not have guessed exactly what sort of vision would be created inside that sphere. ("Hell of Mirrors" 122; NBZ: 264-265)

This is by no means a trivial problem. In preparation for this analysis, I was unable to locate the problem in any math texts, as far as my survey went. I posed the problem to a professional engineer, and a professional mathematician. The engineer called back a week later and said he could not get off the ground with the problem (might I provide him with some assumptions?), and the mathematician did not get back to me. The following problem is posed on a web-based mathematics bulletin board, where it is one of only nine problems out of 200 to receive a designation as "difficult," the only higher designation being "research level problem."

> Subject: Inside a Spherical Mirror
> Date: 25 Feb 2000, 19:43:31 - 0500
> If a person were to stand (holding a small light) inside a sphere whose inner surface was a mirror and looked up, what would they see?

Likewise, what would a person standing inside a similarly mirrored torus (hollow donut) see? [14]

The problem had received no answers in the year and a half since it was posted.

The term "spherical mirror" is actually quite common in engineering and mathematics texts, however it designates there problems involving the fabrication or reflective qualities of a small section of a spherical mirror.[15] Edogawa is well aware of this simpler problem:

> The only thing we know is the reflection cast by a concave mirror, which is only one section of a spherical whole. It is a monstrously huge magnification. But who could possibly imagine what the result would be when one is wrapped up in a complete succession of concave mirrors? (122; NBZ: 265)

The key relation in the discussion of a mirror-observer system is the shifting relation of the observer, a background which is independent, and the source of the light-rays that constitute the image. There are all sorts of considerations involving the location of the object, whether rays are coming in parallel to the optical axis or not, location of the focal point, the geometric center, etc., and the location, size and orientation of the image changes as these parameters shift in relation to each other. But the relevant point for this discussion is that all such diagrams have a nebulous "region A" in the background that does not belong to the system, and from which the rays originate.

If you imagine yourself inside a spherical mirror, you can see that from a technical point of view, what a spherical mirror does is defeat the shifting of the background, and of the angle of incidence, by which one differentiates oneself from the visual field. In a normal flat mirror, any movement of the observer or the object producing the image creates both a shift in the background, and the angle of incidence. A concave mirror seeks to minimize or defeat *the shifting angle of incidence* and attendant distortion at the edge of the field, however the shifting relation to the background, which is independent of the observer-mirror system, remains. As an observer turns in front

14. <http://www.cut-the-knot.com/exchange/mirrorSphere.html> Posted 25 Feb 2000.

15. A discussion can be found in any basic book on optics. See Eshbach, *Handbook of Engineering Fundamentals*, 9-10,11.

of the mirror, images shift in the background, as if one is scanning one's eyes across a painting.

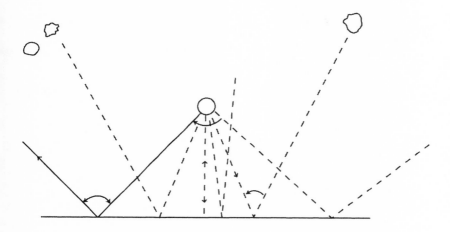

Fig. 9 Flat Mirror: Shifting Angle of Incidence

A spherical mirror defeats both these cues that allow one to distinguish oneself from the visual field. Again, imagine yourself turning slowly in front of a flat mirror. Or better yet, stand in front of a mirror and turn slowly. Watch the background change. Now imagine yourself turning inside a spherical mirror. Because the background itself is composed of reflections from the mirror, instead of constituting an independent source of optical input, the background will be subject to an infinite regress.

In all diagrams you will find of the "spherical mirror" (i.e., concave mirror) problem, you will find a common point: the rays of light reflected in the mirror come from a source that is independent of the mirror-observer system (See Fig. 11).

The nebulous region A, which is unspecified but conditions the system, is of course the object of observation, like, for example, a distant star. The problem is premised, as we would expect, on a separation between the subject and object of vision. The special case of looking at yourself in a mirror fits with this conventionally because there is a premise of a split between mind and body, where the mind occupies the role of the observing eye, and the body occupies the role of the object. What sets the problem of an actual sphere off is that it eliminates this nebulous "region A" which conditions the problem. What happens the moment you close the door and remove even a pinhole of external light (that enables, for example the black box problem), is that the source of the image loses its independence and becomes *internal* to the system. This is what makes it so hard to visualize the problem on paper. We aren't allowed to posit this region A as the source of the image, and get all tangled up. When one closes the sphere, as the logical conclusion of the problem of a "complete perspective," one eliminates this nebulous premise of a "region A" as the external source of the image, and the problem becomes, precisely, self-referential.

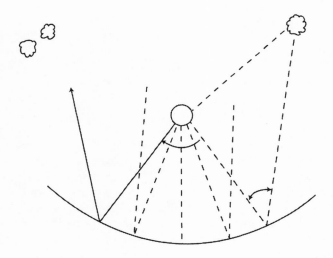

Fig. 10 Concave Mirror: Reduction in Shift in Angle of Incidence

Now, as to what specifically the protagonist *sees* when he entered that mirror and closed the hatch, recall that what produces the distortions when you move a flat mirror is the shifting relation of the perceiving subject to what is behind them, while with a spherical mirror section it is the shifting of relations at the optical axis. One might be tempted to imagine that what would be produced inside a complete spherical mirror would be a funhouse mirror-type of distortion in which the head expands and is curved back in on itself. Or a faceted distortion, as in a compound eye. But the problem is sui generis from the case of a partial mirror, more radical than these manageable distortions. Inside a spherical mirror, no matter how one

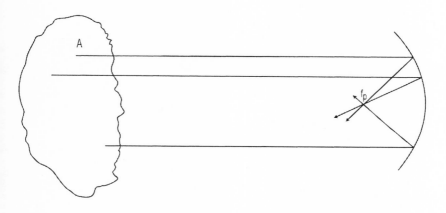

Fig. 11 The "Spherical Mirror" Problem (where spherical mirror denotes a converse or convex section of a mirror, not an actual sphere)

turned the reflection from the rear would stay the same. And because the image source loses its independence and becomes internal to the system, the image would be subject to infinite reflection, with no sink (i.e., no space) to absorb the excess. In such a case, one of two things could happen in the infinite reflection internal to the sphere: First is *cancellation*; whereby all imagery but what is directly in front of the observing eye will echo around and never hit the focal point, which would be like the calm center to a storm. Second is *resonance*; whereby the infinite reflection of images reinforces and amplifies into an infinitely dimensional soup or barrage of unimaginable intensity.

It is difficult to say what the psychological effects of the latter might be, but the former case is highly interesting theoretically. When the observer is displaced from the focal point in the center of the sphere he will see this infinitely dimensional soup. But at odd moments when the eye passes through the center what one will see is a singular representation of what is directly in front of the observer. All other input will diffuse away. But when one considers that, from the point of view of the physiology of the retina, there is a blind spot directly in the line of sight, what one would see at this point would be *nothing*. It is not that you would be blind. You would actually *see* nothing. Pure aporia.[16]

> What terrified me was the fantastic way this sphere kept rolling slowly and haphazardly, as if it were alive. More uncanny, however, was the strange noise that echoed faintly from the interior of the ball—it was a laugh, a spine-chilling laugh that seemed to come from the throat of a creature from some other world. (116; NBZ: 263)

This, at least, is my hypothesis.

The voyeurism, the Gothic nonsense, that is all a cliche. What is really shocking is the mathematics. The idea of zero has existed in India since antiquity as a way to express the idea of not moving a bead in an abacus column, that is to say, as a practical problem. However, its introduction in the West in the 12th century is said to have caused great panic. (Karatani 1995, 40-41) That is to say, Western philosophers immediately recognized the terrifying implications of the mathematical singularity. One can imagine as Tanuma rolls around inside the sphere, testing the effects, that there would be a moment when one eye passed directly through the focal point. At this point, he is infinite, the origin of the system, he becomes God, which in the Gothic imagination is a Hell, an Infinite Terror.

The Romantic Lineage

In this Edogawa reveals a particular Romantic lineage. He has figured the Romantic problem of Kant's "intellectual intuition." In a survey of continental philosophy, Roger Scruton writes that the intellectual intuition, a sort of quick glimpse of the thing-in-itself beyond appearance is an "insignificant appendage" to the Kantian system.

> We could know the thing in itself only if we could obtain an 'intellectual intuition': something which to us is inconceivable. (To have an intellectual intuition is to know the world as God

---

16. Schematically, there are two oppositions at stake in this hell of mirrors:
1) Mortal   —   Divine
2) empirical   —   transcendental
And this leads to two possibilities in interpreting this singularity:
a- interference =   nothing   =   he becomes transcendental subject
b- resonance   =   infinity =   he becomes God

knows it; with an immediate grasp of its totality, and from no finite point of view.) (Scruton 1994, 196)

This is what happens inside the spherical mirror. The protagonist experiences an immediate grasp of its totality, from no finite point of view. And he emerges insane. Scruton continues in a way that situates Ranpo's vision in a particular lineage of romanticism:

> The concept of the intellectual intuition is, I believe, an insignificant appendage to the Kantian system. But it was received by Kant's immediate followers with rapturous applause as the clue to "any future metaphysics." Intellectual intuition became . . . the Holy Grail of German philosophy: to obtain it would be to reach the perspective of the Creator, the coveted view of the whole of things. (196)

Ranpo constructs an exact visual analogue for the strange point to which Kant's philosophy was led but to which it had no inclination to go. Kant calls it God, something which is to us inconceivable. Ranpo calls it Hell and insanity, God's position, appropriately inverted by the mirror.

Is it really necessary to bring all this about Kant into an essay about Japanese literature? The question comes up of whether this is not capriciously reading into a context not really concerned with Western thinkers. Indeed, it is probably easy to say, at some converse logical, epistemological or metaphysical level, that *all* modern writers are talking about Kant. Just as one can read any story with foregrounding of vision and obsessive refinement of the scopic field in terms of the Lacanian Gaze, and any story with a mirror in terms of lack. However, it is possible to raise some points of context to buttress the reading of the logic of the spherical mirror. There is no question that Kant was part of the Tokyo intellectual milieu in the early twentieth century. We have seen the introduction and immediate grasp by Japanese students of Kant in the 1880s through Fenollosa's lectures, and by Edogawa's time a passionate engagement with modern philosophy was de rigueur for any "literary youth" in Tokyo and Sakaguchi Ango recalls *Dekanshō* as a buzzword from his youth in the 1930s meaning "Descartes/Kant/Schopenhauer" (Roden 1980). It is reasonable to presuppose an acquaintance with Kant by Edogawa, just as we can assume he knew optics.

To give an idea of the way this sort of concern was discussed in the Tokyo literary magazines, one can look at a lengthy essay published in 1907 in the journal *Taiyō* by Saitō Nonohito (1878-1909), younger brother of the leading Meiji critic Takayama Chōgyu. Entitled, "Izumi Kyōka and Romanticism."[17] The essay traces the lineage of romanticism to Kant, whose distinction between the phenomena and the thing-in-itself in the *Critique of Pure Reason* is said to institute the modern world view by undermining the enlightenment faith in the ability of reason to grasp the totality of things in the world. The subsequent development of romantic thought through Fichte, then, is an attempt, through a notion of intuited knowledge of the thing-in-itself, to avoid the specifically modern despair that this inability to know must generate:

> But can there be anything unhappier for us than the impossibility of philosophical knowledge of the universal? We believe in gods through faith, we struggle ethically, we contemplate and diligently hone our philosophy, in each of these in fact is sought a negotiation and accommodation between this thing called "I" and the universal. We cannot tolerate the idea that we are like a stone strewn aimlessly on the wide plain, rolling along in the barren universal. Did we not say that the universal is unattainable? We cannot know what, or even where is this universal that would submerge and dissolve us. The goal of this journey of life vanishes before us like a phantom. Faced with this we fall inevitably in lamentation. We despair, we perish in madness and indignation. The irrevocable decision (*tetsuan*) of Kant's philosophy is indeed a death sentence. (Saito 1907, *Tensai* version, 3-4)

Saitō seems puzzled by the fact that Kant, nonetheless, did not himself commit suicide and attributes this to his ability to pull back through faith and construct an ethical understanding of the world. The suicide of von Kleist (1777-1811), then, is offered as an example where a soul, lacking Kant's masculine sense of duty and "Prussian disposition," in the face of a principle that granted no space for life, went on a "journey of discovery" to death. In a climax worthy of the best detective fiction Saitō concludes, "nobody killed von Kleist, he was killed by Kant's *Critique of Pure Reason*." The mind races to construct the film version. The epilogue continues:

---

17. Saitō Nobusaku (Nonohito is his pen name). "Kyōka to Romanchikku." *Taiyō* S40: 9 (1907). Reprinted in a *Shinshōsetsu* supplementary volume *Tensai: Izumi Kyōka.*

This philosophy of the unattainability of the universal cannot be born by us and in order to save us from madness and death, it is inevitable that something would arise of itself to make good the insufficiency in Kant's philosophy. Fichte, that is to say, romanticism is that something. Here Kant's inherent dualism becomes aggressive. Here this thing called "I" is amplified to the extreme, until finally it swallows the universal at a gulp. The universal itself, in other words, designated by Kant as unknowable, is nothing other than this "I." Here Kant's philosophy becomes absolutely subjective, the guarantee of the self's unlimited and absolute freedom and independence. (Saito 1907, 4)

Edogawa graduated from Waseda University in 1916, already decided on a career as a writer, and would certainly have been aware of Saito. Though I cannot verify that he read this, it is almost irresistible to conclude that he has followed this charismatic young theorist of "genius," and engineered in his spherical hell the "I" amplified to the extreme, swallowing up the universal at a gulp.

Deconstruction = Gödel = Hell of Mirrors
Introducing Gödel as a figure for mathematical formalization is a way to make the problematic of deconstruction appear outside the field of literary criticism. Just as Marx introduced money into the classical economical analysis of commodities, Gödel introduces natural numbers into the system of mathematics that takes them as its condition of possibility. In this way both thinkers introduce the self-referential paradox into stable, grounded systems. In Edogawa's literate hands, the quest for the perfect perspective becomes another example, like zero, like Gödel's incompleteness, like money for Marx, where the system turns in on itself, and produces God. That is to say, it is a deconstruction, in Karatani's sense, the indeterminacy that appears when the logic of a formal system is pursued to its limit, the exterior conditioning it is lost, and a self-referential loop appears. It is what made Frege nearly break down when Russell pointed out the fatal paradox. A transcendental perspective, or archimedian perspective, is often imagined to be way up *high* or very far away. That one would be looking at things from a height far removed. God is imagined to reside in the sky. But that is a perspective as much as any other. In "Hell of Mirrors" Edogawa quite rigorously sets up the question of the transcendental perspective, and by pursuing the protocols of a perspectival logic proposes a different solution: the answer lies inside.

This leads Edogawa to a deconstructive (not irrational) conclusion to a genre typically characterized as mystical.

5 Maeda Ai: Topology and the Discourse of Complexity in Japanese Literary Criticism

> I believe that not trusting jargon is the mark of critical spirit.
> —Kobayashi Hideo, "On Theory and Practice" (1995, 108)

Any engagement with Japanese literary criticism over the last 15 years is likely to have one into contact with some cognate of the word "topology." A volume called *Hōhō toshite no kyōkai* (Boundary as Method, 1991) refers the reader to a "transformation of topologies" in the Meiji period (toporojii no henyō), while an interview in the opinion journal *Gunzō* is called "A Topology of the Novel" (Shōsetsu no toporojii). The preface to a collection of short stories by a Meiji writer calls attention to his "topology" in literary history ("Izumi Kyōka no bungakuteki isō"), while in Karatani Kojin's *Origins of Modern Japanese Literature* (1993), the set of assumptions Masaoka Shiki held about the relation of writing and voice is said to bring into view a "topology" on which he stands. (Gotō, 1995; Karatani, 1993, 60; Yoshimi, 1991; Muramatsu, 1983)[1]

This is not to say that topology constitutes a keyword or major trend in Japanese literary theory. Indeed, the word is seldom further elaborated, and usually does not interrupt the flow of the argument.

---

1. Karatani uses *isō* in the original. Cf. the 1986 Parco Department Store pamphlet, "Miru koto no toporojii," by Itō Shunji, collected *in Imeeji—Ways of Seeing: Shikaku to media*, (Tokyo: Parco Shuppan, 1986).

Rather, it seems to impart a nuance or significance assumed to be shared by the writers and readers of Japanese literary criticism. Standing on a topology seems to be different from standing on, say, a subway platform, in an unspecified but well understood way. *Nelson's* character dictionary not very helpfully yields for *isō*: "phase [in science]," but the connotations of the term are clear enough that skimming over it with a vague spatial association will leave the net of signification intact. Yet, as one makes one's way to the *Kojien*[2] level over the course of several encounters it becomes clear that all of the several Japanese terms that translate as topology, including *isō* (位相), *isōron* (位相論), *isōkūkanron* (位相空間論), and the phonetically rendered *toporojii* (トポロジー), point insistently to a much more specific connotation: a powerful branch of mathematics called topology. A vague association with topography or topos won't do here anymore, and the reader who has yet to be initiated to its mysteries may be moved to ask, what is topology? How does it help us understand modern Japanese literature?

One may argue that referring to a text's "topologies" is a harmless rhetorical ploy to give the appearance of rigor, something on the order of the pedantic use of "matrix" in English. Or one may point out that the fact that a term *gestures* to something very large and powerful is still precisely compatible with its functioning as jargon. This is to argue that belaboring the question is excessively pedantic or bad form, and points to the tolerance we've developed to buzzwords in critical discourse. If individual critics seem to use the term with little awareness of its specifically mathematical connotations, though, I would argue that the appearance of *iso* and its cognates as a term in Japanese literary criticism is anything but casual, and at its most general level can help raise questions about how institutional configurations can affect the relation between the sciences and the humanities.

Van Rees (1983) notes that a claim of scientificity is a powerful weapon for scholars of literature. Citations of topology in literary criticism serve an important rhetorical function in this respect, given topology's prestigious status as a fundamental, and relatively new field of mathematics. He further suggests that the use of specialized terminology by academic critics tends to point back to the critic who coined the term, or made it famous, and can imply a set of premises.

---

2. The *Kojien* is a standard unabridged dictionary of Japanese.

[Subsequent critics] will be all the more inclined to adopt not only the proposed terminology but also the premises implicit in these terms, if a critic applies this framework in [the] comprehensive analysis of a famous masterpiece. (Van Rees, 1983, 401)

As it turns out, the superficial citation of topology in Japanese criticism gestures back to a remarkably thorough discussion of the problem by the Japanese critic Maeda Ai in the late 1970s, as a part of the project of his influential *Toshi kūkan no naka no bungaku* (Literature in the City Space, 1982). The sixty page introduction to this work, an ambitious attempt to theorize the negotiation between literary representation and lived city space, is perhaps the most systematic and thorough discussion of topology in relation to literature in any language.

The Larger Problem: Complexity
Though topology is not that familiar in North American criticism,[3] it is worth taking a look at the engagement of Japanese criticism with topology because it bears in interesting ways on a key problem in contemporary literary theory: the relation of its critiques of determinacy, objectivity, and the status of referential reality to the counter-intuitive developments of twentieth century math and science (*Stanford Humanities Review* 1994; *SAQ* 1995; Peterfreund, 1990; Plotnitsky, 1995). This sense of a convergence of widely separated fields over the question of indeterminacy operates in both Japan and the United States, exemplified most recently in the simultaneous emergence of an enthusiasm among the general intellectual community for "complexity" as a paradigm for scientific knowledge. The following discussion of the appropriation of complexity, of which the discussion of topology is a part, plays over two axes: the sciences and a theoretically-informed humanities, and their situation in Japanese and North American universities.

Theory of complex systems, or *fukuzatsukei* in Japanese, names a set of explanatory strategies, such as chaos, catastrophe, fractal geometry, affordance, bifurcation, strange attractors, and Poincaré sections that employ a modern treatment of non-linear differential equations or continuous non-differentiable functions to embrace a certain ir-

---

3. The most sustained engagement of topology in American or European criticism is probably the idiosyncratic work of Michel Serres. For an overview See Serres (1995), especially the second section on "Method" where he sketches the seminal importance of topology in his notion of contemporaneous time.

reducible complexity in dynamic systems. From physics and the geometry of nature, to weather prediction, population dynamics, and the study of human cognition, complexity-based theories are being debated as an elegant and versatile way of approaching problems that do not admit of description by the smooth calculus of exponential and periodic functions. If Tzvetan Todorov could define science as recently as 1985 as, "consist(ing) in the effort to replace chaos with order," this can only be read ironically today as none other than "chaos" emerges as a competing paradigm for knowledge in the sciences.[4]

The appropriation of complexity-based theories within the humanities in North America has tended toward piecemeal citation of one strategy or another, and can be seen to have received its impetus and shape through two authorizing acts. First is the valorization of the contingencies and accident of metonymy over the organic necessity of metaphor by deMan in the 1970s, which can be said to have instituted a critical environment where attempts in science to grasp the discontinuous, the irregular, the contingent would present themselves as irresistible authorizing metaphors (Guillory, 1993, 176-267). A second and more specific determination is Lyotard's well-known endorsement in Section 13 of *The Postmodern Condition* of "Postmodern Science as the Search for Instabilities." Here Lyotard posited a series of disciplines forging a new, non-deterministic mode of science based on the insight, deferred for the history of modern science, that continuously differentiable functions are the exception rather than the rule. His citation of particle physics, fractals and catastrophe theory prefigured the interest in complexity in the United States and, with the addition of chaos, the shape it would take (Lyotard 1983, 53-60). The resulting enthusiasm for fractals and chaos and strange attractors in North American criticism though has often seemed to go little beyond the cover art, the science a stepping-off point for ungoverned metaphorical speculation (Johnson 1996).

Intoxicated by lurid Greek names like chaos and catastrophe, the appropriation of complexity in North American humanities has left itself open to the charge of over-dramatization. Virtually all

---

4. Todorov (1985), 370-380. See Mandelbrot (1983) and Holden (1986) for classic accounts of the move to complexity. For more popular accounts see Waldrop (1992), published in Japanese as *Fukuzatsukei* (Tokyo: Shinchosha, 1995) and Ruelle (1992). Freeman and Skarda (1987) is illuminating for the contentious discussion by a variety of specialists of the question of metaphor.

complexity-based approaches share the less than radical sense that behavior that is chaotic, discontinuous, or unpredictable at the local level, is often incorporated in dynamic systems that are regular, robust and predictable at the statistical level. That is to say, complexity is valuable to science because it works in understanding dynamic systems, its efficacy argued in terms of standard scientific virtues such as economy, adequacy of description, predictive power and ability to suggest new lines of research (Davis 1994, 116-119). *Higher Superstition* (1994) consistently zeroes in on this gap between the excited advocacy of a radical rhetoric in the humanities and the ongoing practice of normal science: "If one insists on calling the development of chaos theory a 'paradigm shift' in the Kuhnian sense of the term, it probably does no harm, as long as it is kept in mind that within the scientific community there is not much sense of foundations being overturned" (Gross and Levitt 1994, 94). This divergence between the understanding of science that circulates in the sciences and the understanding that circulates in the humanities has produced an institutional fact of life in North America called the "science wars," and what should be a fascinating point of articulation has become a point at which the divide between the humanities and the sciences becomes painfully evident. This is why I maintain that the relation of literature to math and science is a problem for literary criticism today.

The appropriation of complexity-based theories in Japan differs in two ways. First, in contrast to the largely academic discussion in the United States, the enthusiasm for *fukuzatsukei* in Japan can be said to be a mass phenomenon. Albert Einstein once lamented that efforts to communicate the implications of quantum physics and relativity theory to a wider audience had been made to wait until well after both were safely established, "limiting and reducing the body of knowledge on the subject to a small, privileged group, crushing the philosophical spirit of the people and leading to the gravest spiritual impoverishment" (Quoted in Virilio, 1994, 23). Whether one takes it as evidence of philosophical spirit or a higher intensity of intellectual fashion, one need only pick up the Tokyo weekly *Shūkan Diamond* to find an uncompromising debate on the epistemic significance of complexity by several of the leading figures in the field offered for the consideration of the general reader. Kodansha, hallowed mass-fiction publisher and leading edge in the postmodern attenuation of the distinction between serious thought and entertainment in Japan has an introductory volume out called *What is Complexity?*, while a ground-up revision of Tokyo University's introductory curriculum includes a section called "What is Chaos?" *Gendai shisō*, a journal of contemporary thought which occupies a position roughly similar to *Critical Inquiry* in English-language criticism, has devoted three separate special issues in the last three years to complexity-based issues.[5] Hence an appropriation and popularization proceeds apace, as Einstein thought fitting, even before the concept has been stabilized in the sciences.

A second point of difference is that the Japanese appropriation of complexity seems to be free of the "science wars" that mar the attempt to articulate a relation between science and the humanities in North America. This is likely less a function of cultural difference than of the common background for intellectuals, specifically a level of math and science literacy that enables an undiluted and mutually informed presentation of the concerns of science.

This is evident when one compares patterns of participation. Special journal issues and books in North America that purport to explore the relation between contemporary theory and the discourse of science, despite gestures to interdisciplinarity tend to be the province of humanities scholars.[6] The resulting discussion of complexity-based theories tends to be metaphorical, seizing on images and quick to dispense with the mathematics that brings scientists to speculate about something like "complexity" in the first place. Such publications in Japan, by contrast, despite their orientation to the humanities consistently bring in practicing scientists and mathematicians to introduce scientific and mathematical topics. In the *Gendai Shisō* special issue on Chaos, for example, the most frequent contributors, in articles, interviews, and round-table discussions were not cultural studies practitioners, but physicists. That cultural critics and physicists in Japan can sit down across the table and discuss the difference, for example, between

---

5. Yoshinaga (1996) is an accessible celebration, while Kaneko (1995) is a superb discussion of chaos in its relation to determinacy and takes a much more cautious view of the applicability outside of physics and math. Special periodical issues include *Shûkan Diamond* (Nov. 2, 1996); *Gendai Shisō* 22.6 (Chaos—Episteme of Complexity, 1994); *Gendai Shisō* 24.13 (Complexity, 1996); and *Gendai Shisō* 25.2 (The Affordance Attitude—An Ecology of Complexity, 1997).

6. One may refer to the contributor biographies in Peterfreund (1990), special issues in *SAQ* and *Stanford Humanities Review* cited above, and the journal *Configurations*, publishing organ for the Society for Literature and Science. An exception in the *SAQ* special issue is mathematician Arkady Plotnitsky. His "Complementarity, Idealization, and the Limits of Classical Conceptions of Reality," (1995), 527-570, is the best discussion of the relation of quantum mechanics to poststructuralist undecidability available.

a metaphorical and mathematical conception of chaos, is in part a product of the math literacy enforced by the examination system. With everyone in possession of the "passport," critical appropriation of science by the humanities in Japan is less likely to leave out the math and caricature the concerns of science.

Conversely, the politics of criticizing the enlightenment project implicit in postclassical math and science carries a different valence in Japan, leading to a willingness among scientists to consider its significance in the wider field of culture. The infrastructure of Japanese science was set in place during the zenith of colonial expansion in the late nineteenth century, and the experience of being comprehended by the institution of western science and culture lends to the discourse of science in Japan an ambivalent relation to its certainties, its hierarchical relations and pretensions to totality. That is to say, the notion that the world is fundamentally complex, that all developments will not ultimately be subsumed under a more general model carries a force counter-intuitive to the logic of modernity. Ambiguously positioned in the narrative of modernity, Japan is a place where working physicists talk about "the écriture of complexity," and situate it in an explicitly postcolonial context:

> It is not simply the appearance of computers that excited the interest in chaos among us physicists. I think it's safe to say that, somewhere we all had a vague sense that the modalities of motion underwritten by what we call the laws of physics weren't all going to yield to explanation by the basic motions, like the periodic movement of a pendulum or the elliptical motion of the planets. This was always in the back of our minds. It's not unreasonable to say that (chaos) is an expression within physics of a kind of unformulated dissatisfaction with the way of viewing things in modernity, where any particular object is apprehended as susceptible to being governed and controlled. (Ikeda 1994, 160)

If a relatively high level of math-literacy among scholars in the humanities provides a common ground for discussion, workers in science in Japan are more likely to reciprocate by being conversant with the stakes and relevant arguments of postmodernist theory, and less reluctant to see in its drive to interrogate the political, economic, scientific and military adventure of western modernity a common project.[7]

---

7. Ikeda goes on to connect the question of chaos quite explicitly with postcolo-

This brings us back to Todorov. The characterization of science for an American audience as consisting in a quasi-ethical opposition of order and *chaos* is of course a caricature of the notion of scientific reductionism, which is perhaps more finely expressed as a problem of order and difference. We may revisit Feynman to oppose to Todorov's caricature an unapologetic representative of scientific thinking.

It is always as complicated as that, no matter where it is. Curiosity demands that we ask questions, that we try to put things together and try to understand this multitude of aspects as perhaps resulting from the action of a relatively small number of elemental things and forces acting in an infinite variety of combinations. . . . In this way we try to gradually analyze all things, to put together things which at first sight look different, with the hope that we may be able to *reduce* the number of *different* things and thereby understand them better.[8]

The institutional differences I have argued above reveal themselves at the formal level, in the way the subject is grasped. The piecemeal appropriation in North America of individual strategies like chaos, fractals, and catastrophe represents the formal expression of an institutional split and the projection of millenarian fantasies by the humanities onto an indifferent science. Proceeding from an apprehension of the project of science in the melodramatic terms of an opposition between order and chaos, academic cultural criticism in the United States indulges in an uninformed fantasy of a revolutionary overturn-

---

nialism: "Something I heard a long time ago from a polymath friend of mine about disciplines like ethnography or history jarred a sense of recognition in me. There are things in the purview of these fields that just don't come to the surface with a fixed or universal perspective, and the crucial thing is to find a way of apprehending things that can see the ceaseless accrual of new meaning that infiltrates by these objects that don't register. Look at the way the European universe is ceaselessly eroded from the exterior by the Africa and Asia they once colonized. Immigration from Africa and Asia embodies this today . . . . The view of the world (*mono no mikata*) is ceaselessly eroded from without. In other words Asia, or Africa, they pierce the interior. Grasping that kind of process in a discipline like ethnology, or philosophy, means that the central framework has to change. This kind of knowledge has been arising simultaneously at the cutting edge of every discipline."

8. Feynman (1995), 23-24, italics in original Cf. Kant, in the Preface to the 2nd ed., *The Critique of Pure Reason*, the central claim of which is to take metaphysics from the realm of scholastic disputation and put it on a firm scientific basis: "It is extremely advantageous to be able to bring a number of investigations under the formula of a single problem."

ing of the rational, orderly processes of modernity. The synthesis in Japan of these strategies under the rubric of "complex systems" or *fukuzatsukei*, on the other hand, represents the tolerance for ambiguity enabled by a shared background, and a nuanced understanding by the humanities that what is at stake here is not the abandonment of determinism or the dissolution of modernity into chaos, but the question of *reductionism* as a mode of knowledge. Further, the notion of an attenuation within science of the demands of reductionism would resonate with a well-grounded anxiety about what it means for the comprehending gaze of western modernity to "*reduce* the number of *different* things," conditioning the intuitive affirmation of complexity, and finding ground prepared for popular excitement about the concept. The analytical work by Japanese discourse, though, of reducing all these different strategies to a single elemental move to complexity, is of course an elegant reductivist move.

Topology should form a set with complexity as an object of desire for literary criticism. In fact, given the evident affinity of a topological transformation of space to one of the central tropes in poststructuralism, the deconstruction of inside/outside in relation to the body, it becomes less a question of why this curious discipline pops up in Japanese criticism than why there is no equivalent discussion in English. Further, as a subject whose customary prerequisites are, as the author of a course on general topology ominously observes, "calculus plus a degree of mathematical maturity," (Christie, 1976, Introduction) we would expect Maeda's work on topology to respond to the institutional determinations I have argued effect the overall symmetry in the discussion of literature and science. Indeed, Maeda's work in the 1970s can be seen as an anticipation of the problem of complexity, and the engagement by this critic of topology emerges as not just a topic of intrinsic or historical interest, or a special local concern of Japanese studies, but as a site for thinking through the institutional requirements, limitations and possibilities of exchange in the emerging discourse of complexity.

This chapter will read in detail a key point in the introduction of topology in the long preface to Maeda's *Literature in the City Space*, the generation of "neighborhood" in section two as a mediating concept for readings of Flaubert's *Madame Bovary*, Mori Ogai's "Maihime" (The Dancing Girl, 1893), and Izumi Kyōka's *Teriha Kyōgen* (The Kyōgen Troupe, 1896). Though I will argue that Maeda's project ultimately fails to articulate the mathematical and literary spaces involved, the way it fails brings into sharp relief larger questions of how theory operates in the production of knowledge, and indicates both promise and limitations in the expectation that Japanese criticism is in a privileged position to articulate the postclassical concerns of science and literature.

What is topology?
    Topology can be said to have originated with the speculations of Euler in the eighteenth century, and to have begun to assume its present form with Weierstrass, Kirchoff, and others, who found ways to articulate abstractly the persistence of certain properties of geometric figures (such as dimension, number of edges, number of sides, continuity, etc.) through all manner of bending, twisting, and stretching, that is to say, through all manner of deformation of Euclidean space. Developments by Cantor in set theory in the late nineteenth century, along with the spread of the conviction that the concept of dimension can be extended to non-spatial coordinates, provided the foundation for an abstract conception of topology, and prepared for its exuberant growth in the twentieth century. Now, "like logic and set theory, [topology] is so fundamental that it infiltrates nearly all of mathematics" (Chinn, 1966, 2; Christie, 1976).
    Topology encompasses many different ideas, from the so-called rubber sheet geometry, which is easy to visualize, to highly abstract locale and information theories that admit of no metaphorical understanding. In order to enable us to place Maeda's use it may be helpful to briefly explain four of these ideas, which may be regarded as increasingly abstract stages:[9]
    1. The first idea is topology as *rubber sheet geometry*. It has been said that "topologists are mathematicians who do not know the difference between a cup of coffee and a doughnut." The cup with a handle and the ring doughnut are topologically equivalent because, given a pliant enough material, one can be stretched and formed into the shape of the other. In fact, given the continuity of the digestive tract, the human observing the two objects is also topologically equivalent. The topologically invariant property is the hole (in the handle, in the doughnut, in the person), which persists through any conceivable manipulation that does not tear the figure. Similarly, there is no topological difference between a triangle, a circle, and an irregular closed shape of any sort, because each can be stretched into the shape

---

9. I take this discussion from the excellent introduction to Vickers (1989), i-xi.

of the other, All, however, are different from a figure with a break or intersection.

Fig. 12 Topological Equivalence   Topological Difference

2. The reason that tearing a topological shape alters its topological properties is that it creates new boundaries. Topology as *the study of boundaries* approaches sets in terms of properties like open and closed. Closed sets include all their boundary points, while open sets include none. The concept of neighborhood is used to define the boundaries of a set in a way that incorporates the idea of nearness, or tolerance. A boundary point is one which "however closely you look at it, you can see some neighboring points inside the set, and some outside."

3. With the further abstraction to *the study of open and closed sets*, we lose the ability to understand by the metaphor of geometry. Topological arguments from stage two are translated into an abstract set of points called a *topological space*, which can have any number of dimensions, and is specified in certain ways relating to open and closed sets.

4. In a progression similar to the eventual abstraction of the properties of geometry in quantum descriptions of matter, *locale theory* takes us further into a mathematical world, "forget(ting) even about the points," and by replacing functions such as union and intersection with abstract algebraic operations, manipulates abstract sets of "open sets."

The applicability of topology to sciences other than mathematics is well established. The correspondence with topological concepts and operations is usually effected by an intervening mathematical subject, and occurs in the proof of theorems about the basic structure of the science. In network theory, for example, topological study, mediated by the application of linear graphs has become indispensable to modern engineering (Kim, 1962; Pullen, 1962). The description of networks such as an electrical circuit or traffic flow can be reduced to certain topological variables, and all other variables disregarded. For example, it makes no difference whether the road between two traffic lights is one mile long and straight or two miles long and curved. None of these affects the flow of traffic between the lights. If the road were to fork, however, or another intersection were added, this would change the topological properties of the traffic flow. In the same way in electrical engineering, nodes, paths, and elements signify, while the shape of the circuit in Cartesian space does not. These are but a few of the fields where topology has had a fundamental impact, and the most easy to visualize. Attempts have also been made to apply topology to fields whose suitability to mathematical expression is not as clear, such as the social sciences. Maeda's mobilization of topological analysis in studying modern literature can be seen as such an attempt.

Topology in Literature in the City Space

With this brief exposition of topology on the table, I would like to work through Maeda's generation of a critical construct for reading space through this mediation, and then follow step by step as he carries the resulting theoretical schemata into the domain of literature proper.

"Kūkan no Tekusuto: Tekusuto no Kūkan" (Text as Space: The Space of the Text) is the lengthy theoretical introduction to Maeda's seminal *Literature in the City Space* (1982). Like the main part of the book, it ranges widely over modern Europe and Japan, with richly detailed readings of works as diverse as *Madame Bovary* and Mori Ogai's "Maihime," from Tachiwara Michizo's sonnet "Watakushi no Kaette Kuru no wa" (The Things I Return To) to the 1980 best-seller *Nantonaku Kurisutaru* (Somehow . . . Crystal), from the telephone book to the Tokyo entertainment weekly *Pia*. Unlike the remainder of the book, though, the readings of the introduction are interspersed with explicitly theoretical reflections on the negotiation between city space and literary text, with *Madame Bovary* and "Maihime" taken up in the context of the phenomenological theories of Georges Poulet and Roman Ingarden, and Tachiwara Michizo and *Nantonaku Kurisutaru* read in conjunction with Heidegger's notion of the *thing*.

Running through the entire introduction is an ambitious and highly technical exposition of the theories of topological analysis, from the definition of limits in metric space, to Hausdorff spaces and discrete and indiscrete topologies. Each of the five sections of the introduction explores a different facet of topology. These expositions include extensive use of mathematical notation and diagrams, and are followed by elucidation through the literary works under consideration.

Before bringing our attention to bear on Section Two it will be worthwhile to briefly discuss Section One, since topological terminology makes its first appearance there, and the discussion of neighborhood proceeds from it. In section one, Maeda takes up George Poulet's phenomenological reading of *Madame Bovary* wherein Poulet interprets the project of Flaubert's fiction as that of instituting, through the metaphor of the circle, a new, more convincing, and properly phenomenological "way of presenting the relations between the mind and all surrounding reality" (Poulet, 1966, 263). Here the text's interior space begins to live for the reader of literature through possession of a circle defined by the narrator's viewpoint, or the gaze of a character turning to the things around them.

Topological terminology first appears when Maeda is criticizing the persistence in Poulet's work of the Cartesian *cogito* along with Husserl's intentionality, and we may associate Maeda's mobilization of topology with his concern to escape this framework in dealing with space and text. After noting the breathtaking scope of Poulet's argument in *Metamorphoses of the Circle*, from Parmenides to Eliot, Maeda poses a question:

> We may ask, however, what motivates Poulet's conception which, while narrating the various topologies (*isō*) of "metamorphosis," never separates from the topos of the circle. (Maeda 1982, 8)

Poulet's circle is from then on called a "topos," which may be the more neutral "space," but given that the first instance explicitly refers to *isō*, appears to imply that Poulet's circle, persisting through 2000 years of transformation, is a topological figure. Reference to Poulet corroborates:

> By "metamorphoses of the circle" one must therefore understand, not the metamorphosis of a form by definition non-metamorphosable, but the changes of meaning to which it has never ceased to adapt itself in the human mind. (Poulet 1966, vii)

Persistence through change is a primary property of topological figures, but we may remember that it is precisely geometric shape that can be changed while preserving topological equivalence. There is no topological difference between the circle, and triangles, pentagons, or any other simple closed curve. Poulet though, is clearly talking about "the circle that Euclid describes," which, because of its "simplicity, its perfection, its ceaseless universal application . . . traverses time without being affected by it" (Poulet 1966, vi). It is not clear, then, why Poulet's circle, which remains a circle through all metamorphoses of meaning, should be discussed in the language of topology, and the matter is dropped.

Maeda then moves to a discussion of Roman Ingarden's *The Literary Work of Art* (1973), drawing from Ingarden's phenomenological analysis of literature the notion that the act of reading involves the assumption of a "center of orientation" which is the "here" of the represented space of the text. This notion is seen as corresponding to the Husserlian "zero point" by which lived space is oriented by the perceiving body. For Ingarden, when we enter the represented space of the literary text, we are shown aspects of things that imply a center of orientation. This center may move with and correspond to, a narrator, to one or several characters in succession, or it may be disembodied, but it always represents the point from which the people, animals and things of the represented space are "perceived." This will become important later as Maeda introduces the topological concept of neighborhood. We may note now that another topological term makes an appearance here. In discussing an early scene from *Madame Bovary*, Maeda describes the list of things that enter Charles' field of vision as he sees Emma for the first time (window, sunbeams, stone tiles, the fly in the cider cup, and finally Emma's bare shoulder) as a mathematical set (*shūgō*) of *things*. The notion of "set" again foreshadows the coming discussion of topology, while the emphasis marks beside *mono* look forward to the discussion of Heidegger in section four. My own understanding of "thing" in Heidegger is that it is meant in part to call into question the scientific abstraction of qualities by pointing to a notion of "thing as thing." Maeda seems to have a similar idea in mind in section four where he contrasts the *things* that populate Tachiwara's sonnet "The Things I return to" with the brand name commodities that make up the represented space of *Nantonaku Kurisutaru*. One must again wonder how this will be related to a branch of mathematics that "can't tell the difference between a cup of coffee and a donut," or Heidegger's jug for that matter (Heidegger, 1971, 163-186).

What we find is that Maeda has seeded section one with topological terms whose presence, while schematically plausible, does not invite close scrutiny. It is perhaps more important at this point, though, to concentrate on the notion he sets up of Ingarden's center of orientation as effecting for the reader a kind of zero point or "here" in the represented space of the text. This will be crucial for his effort to link the literary text with topology in the more thorough discussion of section two.

Section two begins with a discussion of the Tokyo entertainment weekly *Pia*, a thick, glossy publication filled with information on things to do in Tokyo. *Pia* is discussed in light of Michel Butor's provocative suggestion that literature, which we read from cover to cover, only represents a small portion of the texts we encounter, and that we need to pay more attention to phone books. The negotiation between the names in a phonebook, or the information in *Pia*, and the city space, is discussed in terms of sets, between which a mapping function of one-to-one correspondence applies. In other words, there is one and only one phoneline corresponding to each number recorded in the phonebook, just as there is only one movie theater, for example, corresponding to an entry in *Pia's* Cinema section. It is the special characteristic of the literary work to lack this one-to-one correspondence, a definition which will be picked up again in the context of Yuri Lotman in section three.

It is after this discussion that Maeda takes up the question of topology in earnest in the following passage:

> As I suggested when I touched on this before, Ingarden moves toward a definition of the represented space (*naikūkan*) of the literary text as a transposition of the spreading, lived space oriented around the zero point of the I. In place of the I, then, is the center of orientation inside the text, a narrator or a character, and the various things depicted in the text and the space which includes them appear as a space oriented by this narrator or character. The center—circle gestalt developed in Poulet's *Metamorphoses of the Circle*, while structured by a slippage, also corresponds to this schema. One may note that this schema of Poulet and Ingarden bears a resemblance to the concept of *neighborhood*, well known as one of the axioms of the theory of topological space. In fact, if we hypothesize that the literary text is a finite point set, a set of words actualized by the reading process, and select a random sample from a continuous act of reading, we can say that the neighborhood model comprehends the schema of Poulet and Ingarden. Further, we would expect it will be possible to fix the topology of the text as a whole by using the neighborhood model. (17-18)

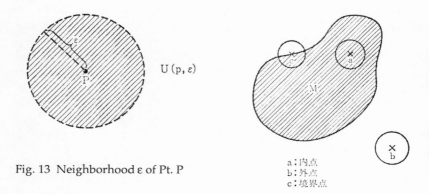

U (p, ε)

Fig. 13 Neighborhood ε of Pt. P

a：内点
b：外点
c：境界点

a = interior point
b = exterior point
c = boundary point

Fig. 14 Partial space M with boundary points
(from Maeda 1982, 19)

I have quoted this passage at length because it is here that Maeda formally introduces his project with respect to topology. The concept of neighborhood is to be used to bracket the phenomenological approach (with its echoes of the *cogito* in Poulet's case), and "fix the topology of the text." It is also worth noting, however, the striking rhetorical shift in the second paragraph. The sudden introduction of mathematical terms like "axiom," "hypothesis," "finite point set," "random sample," and topology, signals a shift, from the language of phenomenology, reassuringly centered on the human perceiver, to the detached precision of the mathematical treatise. This skillful modulation of different rhetorical registers is in fact, one of the principle pleasures of "Text as Space—Space of the Text." Each of the five sections of the introduction moves us up and down this rhetorical rollercoaster (or sine wave), from richly concrete discussion of the everyday things and people by which Tachiwara or *Nantonaku Kuris-*

utaru evoke their represented space, to anthropology or phenomenological exposition, to the authority and serene precision of mathematics, and then back to a literary text with a quarry of new concepts.

Shortly after the passage quoted above, the register shifts into
mathematical exposition proper as Maeda introduces the concept of
neighborhood. To get a sense of the range of rhetorical register in the
text, and on the principle that it is difficult to be more concise than
mathematical notation, I will translate Maeda's exposition in full
(found on pp. 18-20), and draw on it as needed for the discussion
which follows:

1. One of the axioms which fixes the topology of a metric space is
    neighborhood.

2. Given pt. P of metric space (X,d)[10] and positive real number ε, when
    metric space X is of area R^2, we call the interior of the circle of
    radius ε with center at pt. P (the borderless disc) the **neighborhood
    ε of pt. P** (or simply neighborhood), which we write U(P, ε). i.e.

    $$U(P, \varepsilon) = \{x \mid d(P, x) < \varepsilon\}$$

    Further, we call the neighborhood function the set of all the neigh-
    borhoods ε of pt. P as ε moves through the range of all positive real
    numbers, $\{U(P, \varepsilon)\}$, $\varepsilon > 0$. (see Fig. 13)

3. When metric space X is of area R^2, if we take M as a single, given
    partial set of X, a pt. P of X can be classified by the distribution of
    pts. x of M in a neighborhood of P. (Fig. 14)
    a) With regard to a pt. P of X, for an appropriately chosen value of
    ε, when

    $$U(P, \varepsilon) \subset M$$

    holds good, we call P an **interior pt.** of M. We call the set of all
    interior pts. of M the **interior** or **open kernel** of M, designated M^i.

---

10. Metric space X governed by distance function "d," where "d" is a function
which precludes discontinuity and defines a space as metric. Maeda introduces a
number of terms without definition or explanation.

b) We call the interior pts. of the complementary set of M with respect to R^2 the **exterior pts.** of M. i.e., a pt. P is an exterior pt. if there is a neighborhood U(P, ε) which satisfies,

$$U(P, ε) \subset (X\text{-}M)$$

The set of all exterior pts. of M is called the **exterior** of M, designated M^e.

c) Pts of R^2 which are neither interior pts. nor exterior pts. of M are called **boundary pts.** of M. i.e., for a random neighborhood of P, U(P, ε),

$$U(P, ε) \cap M \neq ø, \text{ and also}$$
$$U(P,e) \cap (X\text{-}M) \neq ø, \text{ both hold good.}$$

The set of all boundary pts. of M is called the **boundary** or **frontier** of M, designated M^f.

Mapping into the Domain of Literature: Madame Bovary to Maihime

With this in place we can follow Maeda's argument step by step as he shifts back to the task of literary criticism through contrasting readings of *Madame Bovary*, "Maihime," and Izumi Kyōka's *Teriha Kyōgen*. Since his reading of modern Japanese literature is heavily determined by his conclusions about *Madame Bovary*, I will follow this discussion first.

The equation of Ingarden's "center of orientation = narrator/character" with pt. P, the center of the neighborhood is perhaps the crucial mediating step in Maeda's movement from mathematical concept to literary text. Recalling how the concept of neighborhood allows us to decide where pt. P falls in relation to the boundary of M, as we follow the literary text through its plot sequence in the act of reading, pt. P then would move with the "center of orientation," appearing in one of three possible places, as either an interior, exterior, or a boundary point. Maeda further specifies that Neighborhood ε of pt. P will be fixed by the gaze or thoughts of the narrator or characters, with the result that the things, scenery, and people within its range, as points x of space R^2, will belong to the interior or exterior of subset M. Clearly, the character of the border of subset M is the principle issue involved.

Taking up again the passage from *Madame Bovary* where Charles, coming in from the bright sunlight adjusts his gaze slowly to the inside of Emma's house, Maeda sets up a preliminary model with the house as subset M. Here Charles, pausing in the threshold, and bringing into focus one by one the eaves of the window, the sunlight through the cracks in the wall, the stone tiles, the furniture, the ceiling, the dinner table, the cider cup with a fly in the bottom, the soot-covered stove lid, and Emma's bare shoulder, occupies a boundary point, with the neighborhood defined by his gaze containing points both interior and exterior to the house. Emma, then, centers a neighborhood q, which lies entirely inside the house. It is the tension between Charles' desire to restrict Emma to this scene of chaste domesticity, and the expansion of her own fantasies beyond its borders, which drives the narrative of *Madame Bovary*.

The critical payoff is soon revealed as Maeda expands his analysis to account for the function of the exterior neighborhoods in defining the boundary of subset M, now expanded to include Emma's village life.

Faced with the bourgeois banality of life with Charles, Emma's dissatisfaction grows. Invited unexpectedly to a ball at a local nobleman's country estate, Emma is able to draw considerable attention with her beauty, and she leaves cherishing the memory of a dance with the Viscount. Emma buys maps of Paris and ladies magazines, novels by Balzac and Eugène Sue, and begins to cultivate an image of the city where the Viscount lives. Maeda cites a passage from *Madame Bovary*:

> In the novels of Eugene Sue she studied the way the furniture was arranged, she read Balzac and George Sand, seeking fictional satisfaction for her yearning. The memory of the Viscount always returned as she read. Between him and the imaginary personages she made comparisons. But *the circle of which he was the center gradually widened around him,* and the halo that he bore, drawing away from his head, broadened out beyond, lighting up other dreams.[11]

Maeda has italicized the cue which will allow him to adapt this text to topological analysis as follows:

> The fantasies that fill Emma's heart, fantasies exterior to the house at Tostes, are cast in the direction of distant Paris and centered on the image of her dancing partner, the Viscount. We may introduce the model of the neighborhood here, and we see that the circle of fantasy Emma has drawn with the Viscount at its center,

---

11. Maeda takes the passage from Poulet, 258, italics Maeda's.

contains not a single one of the details that make up her dull, monotonous country life. That's not all, it denies the very existence of Charles. . . . Conversely, everything that was lacking in the village of Tostes is gathered together complete in her fantasies. The gaslight flickering in the wind, a salon hung all over with mirrors, a circle of dresses ruffling across the floor—all these *things* (*mono*), trembling, dissolving into a circle with the Viscount as its center. This is a representative example of a textual space whose topology is fixed by its exterior points. (Maeda, 22)

After struggling to follow the rarefied mathematical reasoning of the previous pages, the reader's breath is almost taken away as the cues of the textual space of the novel drop one by one into the topological schemata Maeda has developed. Having established Emma as centering a neighborhood of points interior to the house in Tostes, Emma's circumscribed village life is selected as subset M. We know that the neighborhood of a point exterior to M will have all of its points exterior, and are invited to see that, indeed, a neighborhood with the Viscount as its center "contains not a single detail" of her dull country life, and conversely contains everything that the village does not. From this Maeda concludes that *Madame Bovary* is a particular kind of text whose representational space is a topology fixed through the clear delineation of interior and exterior.

Having set up a topological model of literary space and established its usefulness in reading a European classic, Maeda marks the transition to the discussion of Japanese literature (a passage iterated in each of the five sections of the introduction) in the following passage:

Here we have [in *Madame Bovary*] a representative example of a textual space whose topology is fixed by its exterior points.

When we consider Japan's modern novel, we find that texts like *Madame Bovary*, which, in terms of set theory rigidly separate their interior M^i from their exterior M^e, are surprisingly rare. Rather, there seem to be a preponderance of texts of the type which, through the skilful evocation of boundary points, hint at the differences between interior M^i and exterior M^e. For example, there is the scene when Toyotaro, the hero of "Maihime," goes to meet Elise . . . (22)

Maeda goes on to cite the famous passage from Mori Ogai's "Maihime" in which the protagonist Toyotaro passes from the monuments and strolling soldiers of central Berlin's *Unter den Linden* to the winding backstreets where he shares a small attic room with his

German lover. The point to be made is that Toyotaro crosses a border here, between two spaces, one the bright, geometric, imperial space of the Kaiser, the other a dark, erotic, labyrinthine space, and that these are skillfully evoked by portraying the protagonist as a figure who *straddles* the spaces. Maeda then extends the argument by taking up Izumi Kyōka's *Teriha Kyōgen* (1896), to suggest that the boundary points at stake in the literary text need not be construed as geographical space. In *Teriha Kyōgen*, an itinerant Noh troupe arrives in a provincial town on the Japan Sea Coast in the Meiji period, and sets up its tent in a vacant lot by a riverbed. The itinerant troupe represents both fascination and fear to the townspeople, and defines a sacred space,[12] from which the children are warned to stay away. The little boy Mitsugi, whose parents both died when he was young, is drawn to this sacred space by the women of the troupe, and again the salient point is that, just as in "Maihime," *Teriha Kyōgen* is a story whose space is oriented by a hero who, because he freely transgresses its boundaries, is defined by neither the interior, exterior, nor boundary sets.

The critical act by Maeda of resolving a textual space into distinct point sets that comprise a topological space evokes the ambivalence of intellectuals outside the western metropolitan centers by highlighting this tendency of protagonists to carry multiple affiliations, straddle boundaries and cross into new spaces. Representative texts of modern Japanese literature emerge in contrast to *Madame Bovary*, and, carried along by the momentum of a well-turned interpretation, one begins to feel that topology can bring its own rigor to the analysis of literary space.

For the sake of this argument, though, one must note that, although in laying out the space of these texts into three separate regions, interior, exterior, and boundary points, he is specifically contrasting these stories with his reading of *Madame Bovary*, he does not do so by reference to the concept of neighborhood. Indeed, except for the vague use of the words "space" (*kūkan*) and "boundary" (*kyōkaiten*), the topological terminology so painstakingly put into place suddenly

---

12. These itinerant troupes, denizens of the dry river beds in the summer, were often part of the outcast class in the Tokugawa Period, and structured a space of taboo for town and village communities. This was a subject of some importance to Kyoka, who has a number of stories about such troupes, including *Giketsu kyōketsu*, on which Mizoguchi's famous movie, *Taki no shiraito* (1933) is based.

drops out, a lapse that Maeda's readers, attuned to minute shifts in rhetorical register, cannot fail to notice.

One might be inclined to ignore this as a loose end. However, encouraged by the thoroughness of Maeda's own exposition, it may be well to follow this thread back to the point where the topological construct last appeared. Tugging a bit at the articulation of Ingarden's phenomenology with the mathematical definition of neighborhood leads one to further revisit a few seemingly trivial discrepancies in its application to *Madame Bovary*. In fact, the argument will collapse on pressing these points.

The first problem to which we are led is the peculiar qualification "appropriate," to the selection of ε (see def. 3-a), specified in the mathematical definition as necessary for the neighborhood (of radius) ε of pt. P to fulfill its function of distinguishing interior, exterior and boundary points. What does "appropriate" mean here? We may recall that, in equating pt. P with Ingarden's center of orientation, Maeda has selected an ε fixed by the gaze or thoughts of the narrator or characters. Though appropriate from a phenomenological standpoint, is there any reason to think that this selection of ε is appropriate for the specific topological function neighborhood is designed to fulfill.

The second problem follows from the first, and can best be brought out in reference to Fig. 14 (repeated below). Glancing at the figure, we may note that it doesn't represent two seemingly reasonable possibilities, namely neighborhoods "d" and "e" (added to Fig. 15) whose center points are not on the border, but which contain both inside and outside points Since there doesn't seem to be any reason why a narrator or character not exactly on the boundary couldn't take in points inside and outside of whatever boundary the critic wished to establish (by, for example, walking through the door, or taking a train to the next city), we must inquire as to why these possibilities aren't represented in the neighborhood diagram. I will argue that the two cases represented by "d" and "e," while in fact being typical of the way oriented space works in literary texts, actually are not meaningful as topological neighborhoods. This is because the "appropriate selection" of ε actually means appropriately small, so that the neighborhood is contained entirely within, or without, space M.

Perhaps the easiest way to understand why the appropriate value of ε is a small value is to ask why one needs the concept of neighborhood to define a boundary in the first place. Just by looking we can see in Fig. 14 that the center of neighborhood "a" is interior to M and the center of "b" is exterior. Why bother with the circle of radius e? Why not simply designate the points, and draw the boundary?

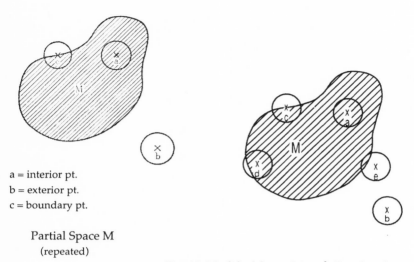

a = interior pt.
b = exterior pt.
c = boundary pt.

Partial Space M
(repeated)

Fig. 15 Modified Space M with Pts. d and e

I would argue that neighborhood is a very specific function, meant to introduce a notion of "nearness" into the limitation of sets. Vickers (1989, 1) notes that the "characteristic of a boundary point of a set is that *however closely you look at it*, you can see some neighboring points inside the set and some outside" (emphasis in original). Hence, the radius ε of a neighborhood can be understood as a threshold of decidability or tolerance of a field, and the two new possibilities ("d" and "e" added to Fig. 15) are in fact, boundary points topologically equivalent to "c." Though we located "d" and "e" slightly off the border of M, our margin of undecidability ε means that we can't tell for sure whether they are inside or outside. The concept of neighborhood, then, is an abstract way to define topological sets whose borders are, not fixed as in Euclidean geometry, but semi-decidable and characterized by this tolerance driven fuzziness.

Maeda is in other sections sensitive to this characteristic of borders. In the following section, for example, he will use Edmund Leach's anthropological notion of liminality as an ambiguous border (Leach, 1976, 81-92) to make a thought-provoking critique of the kind of Aristotelian logic he sees in the work of the formalist Yuri Lotman, which would see no ambiguity in distinguishing inside from outside, structure from chaos, or right from wrong. What is at issue, then, is not Maeda's understanding of topology, but whether the function of neighborhood he develops in great detail, and persuasively deploys in the following section, is tactically neglected in this part of the argument.

Tracing these loose ends into the fabric of the argument leads us to scrutinize a further point: the crucial designation of the Viscount as the center pt. P of an all-encompassing external neighborhood whose points are said to "define the topology of the text space." Having gone through this excruciating math, we are now in a position to ask if this is indeed a neighborhood in the sense Maeda himself has painstakingly defined. It fails the test at three points:

First, the neighborhood seems to have a hole in it. The external neighborhood with the Viscount as center is said to fix the topological space M (life in Tostes) because: a) it contains no details from Emma's boring country life (no points of M), and b) it contains everything that Emma's boring village life does not (all points outside M). The implication is that this single neighborhood is equivalent to the *exterior* of M, and defines the topological space by containing *all* the objects outside (recall that the things and people described in the text are our points). Neighborhoods, though, do not define a topological space by surrounding it. To define the semi-decidable border of a topological figure takes, rather, an infinite number of small neighborhoods very near to the boundary. As a circle, i.e., a simple closed curve, a neighborhood is unable to fulfill the function assigned to it here precisely in a topological sense. A neighborhood is unable to surround a space because that would put *a hole* in it.

Second, the Viscount does not really seem to be a "center of orientation" in Ingarden's sense. We may recall that Ingarden's "narrator or character as center of orientation" has been made to correspond to the neighborhood's center, pt. P. For Ingarden, as one becomes absorbed in a literary text, one loses sight of one's own body as the phenomenological "zero point," and "perceives" aspects of the text's represented space from a center of orientation. Ingarden likens this to walking into the represented space, and walking along with "the represented psychic subjects 'perceiving' this represented space" (Ingarden 1973, 230-231). Center of orientation, then, is a normative reading concept which establishes a "here" from which represented objects in the text (things, animals, people) are seen (touched, heard, imagined). If this is the case, then it is not clear why the Viscount should be the "pt. P = center of orientation" of the fantasies about Paris. Although he may occupy the center of this space, the space is not presented as though perceived by him. The entire fantasy space of Paris, rather, seems to be oriented to Emma, to whom "the memory of the Viscount always *returned* as she read." This memory can return only to the "here" of the center of orientation, to Emma assembling this fantasy space as she reads.

Recasting the pt. P of this neighborhood as Emma leads to two possibilities, neither of which delineates a boundary separating an interior and exterior: a) Emma, though centered in space M, extends her gaze via novels and fantasy outside of Tostes to Paris, hence crosses its borders due to the large value of ε, and is meaningless as a neighborhood. And b) Emma's reading and imagination are interpreted as a kind of nested box of representational or imaginational space, an act of reading represented within a novel, or a fantasy within fiction. In this case, the entire representation is interior to Emma's Tostes, and we still have no exterior neighborhoods to define the space.

Third, the external neighborhood does not seem to be completely external. Granting for a moment Maeda's designation of the Viscount as the center of orientation of the external neighborhood as an attempt to avoid a metaphysical notion of concentric circles which he criticizes in Poulet, one still must note that any neighborhood defined by the Viscount's gaze will include Emma for the seemingly trivial reason that they waltzed together. Again, we have a neighborhood that has an inappropriately large value of ε, reaches into the interior, and is meaningless for defining a topological space M. Or alternately, recalling our discussion of Figs. 2 and 3, we could conclude that both the Viscount *and* Emma are boundary points, with neighborhoods which contain some points inside and some outside of M, though this again suggests a border so fuzzy as to be indiscernible. Thinking about this suggests that the only conceivable "literary neighborhood" which could be exterior (i.e., have an appropriate ε), would have nothing whatsoever to do with the other characters and situations in a text. It would never reach to the represented space M, and would seem disjointed, discrete, perhaps experimental. In Ingarden's metaphor, the reader would be literally dis-oriented, and find herself transported completely outside what would be normatively called the "main" part of the text. In the meeting of the Count and Emma, the "exterior" neighborhood would cross inside to the space of Emma's country life, leading one to question its exteriority, based on Maeda's own exposition of topology.

Well in evidence in the discussion of Ogai and Kyōka's classic work are the qualities that made Maeda resonate with a new generation of critics in the 1980s. In the context of an almost fetishistic focus in postwar academic criticism on author (*sakkaron*) and work (*sakuhinron*) as the locus of literary significance, Maeda's *Literature in the City Space* and *Kindai dokusha no seiritsu* (The Formation of the Modern Reader, 1989), took contemporary readers on a methodological adventure, opening the hothouse of the literary coterie and the writer's private room to such problematics as the development of the city, material changes in reading practices and the circulation and consumption of literature, and liberating critical resources that still resonate today. Typical of all that is best in his work, Maeda's interpretations of "Maihime," and *Teriha Kyōgen* are filled with beautifully selected details, and are well worth reading outside the questions of topology addressed here.

However, the argument contrasting these works with *Madame Bovary* follows in no conceptually consistent way from the critical construct. This is a bit of a problem for the initial hypothesis, which hoped to find a relation between literature and the discourse of math and science in Japanese criticism made rich by a degree of math literacy. What has emerged instead is a disjuncture, between a critical construct whose rhetorical promise is of an analysis of scientific rigor and precision, and the neglect of its implications in the business of practical interpretation.

Crossed Connection: From Schema to Domain

Precisely because of the generosity of Maeda's theoretical exposition, analyzing how the argument goes wrong can be fully as productive for future work as imitating its success. I would argue that the problems with the interpretation of the three stories are neither arbitrary nor the random accrual of errors but point to something else, something besides the theory, that is driving and organizing the critical act. It is useful to distinguish here between the critical failure to be guided by the theory so painstakingly produced, and the more fundamental theoretical problem relating to the construction of a hybrid object whereby literature might be thought through the categories of math or science.

The critical problem lies in the way topological constructs are tactically neglected in producing the interpretation. The chief function of topology seems to be to authorize a contrast between the three texts: *Madame Bovary* on the one hand, and Maihime and *Teriha Kyōgen* on the other. These are further said to be representative texts, Maihime and *Teriha Kyōgen* of modern Japanese literature, and *Madame Bovary* of an unspecified set of "other texts like it." The characteristic claimed for the latter set of texts, that of boundaries rigidly defined and uncrossed, is said to be "surprisingly uncommon" in modern Japanese literature. The footnote to the passage quoted above provides a clue to the identity of the set of which *Madame Bovary* is said to representative by suggesting that this difference may be explainable in terms of the notion of *aida*, which is familiar from the popular discourse on Japanese identity as an idea of sociality opposed to Western individualism by which Japanese social relations are said to be uniquely constructed.

I have already suggested that *Madame Bovary* does not really support the topological claim, that its boundaries are rather always subject to being crossed, its neighborhoods always overlapping. This undoes the contrast in this case. I would further suggest that the tiny overlap, the single dance which joins the neighborhood of Emma and that of the Viscount, and renders them problematic for delineating a topology, is not an accidental feature, which a lot of other texts like *Madame Bovary* will not have. Maeda implies that there is a difference in national culture at stake, but following the implications of topology leads one rather to the question of narrative. Let us suppose that Flaubert, for cultural or national reasons was predisposed to make a text structured by characters completely outside of his heroine's circumscribed country life. He could have chosen as the Viscount any of hundreds of noblemen who would never even know of the existence of Tostes. One need not invoke culturalist explanations of a difference between Japan and Europe in order to address the question of why Flaubert would choose instead the one Viscount who negotiates the border between Paris and Emma's country life, whose neighborhood overlaps with Emma's. This is predictable if we understand *Madame Bovary* to be a narrative text. A certain understanding of narrative would see it as characterized by economy in the presentation of information, and neat causality linking its parts.[13] The logic of the concept of neighborhood cannot distinguish the boundary straddling of Toyotaro and Mitsugi from the causal relays by which the mass of narrative texts maintain continuity. Any narrative of this type will always have

---

13. For a discussion of narrative economy, see J. Hillis Miller, "Narrative," in McLaughlin and Lentricchia (1995); and Bordwell (1985).

this overlap between characters, because causality and economy demand that there be no totally disjointed narrative information.

And yet, most readers would agree with Maeda that the represented space of *Madame Bovary* seems circumscribed and claustrophobic compared to that of "Maihime." What accounts for this difference if not the question of national literatures? I would suggest that had Maeda allowed the study to be guided by the logic of his theory, he might have been led rather to questions of gender.[14]

We may imagine what *Madame Bovary* would be like if the Viscount *really was* the center of orientation. He would perhaps, like Toyotaro, be depicted as moving freely from the bright, official space of the "city of light" across a boundary to the dark, winding space of a provincial night, where, like Toyotaro, he would have a liaison with an erotically charged, lower-class woman. Both Elise of "Maihime" and Emma of *Madame Bovary*, however, must wait in their own circumscribed space. The type of represented space which these two stories share, where the male character moves freely while the woman is made to restrict her desires within the confines of a stifling domestic or local setting, is common in artistic representation across cultures and historical periods, and one may immediately think of examples as diverse as the 11th century *Kagerō Nikki* and the women of 1950s Hollywood melodrama.

Hence what Maeda figures as national or cultural difference emerges through the topological construct rather as a question of narrative or gender. This is not to say that the question of difference among national literatures is not significant. Toyotaro's status as a Japanese man in the heart of nineteenth century imperialist Germany can never correspond with a French aristocrat moving in the milieu he governs. But what emerges from the logic of center of orientation, when applied to the character relationships on which Maeda focuses, is that as men, moving freely over borders and into the circumscribed space in which women are confined, they are strikingly similar. The reason one text seems claustrophobic, while the other is filled with openness and possibility, is that the center of orientation of *Madame Bovary* is a

---

14. Such a destination is not at all unexpected in the context of North American criticism, hence such a critique posed from a position in the North American academy is entirely predictable, and needs to reflect back on itself. However, since my reading does not appear to be at odds with the critical construct one will have to go elsewhere to find the symptomatic stakes of this essay.

female character, and the center of orientation of "Maihime" is a male character.

The question then emerges of what is guiding Maeda's interpretation if not topology? In a study of the practical role of theory in film and literary criticism, David Bordwell constructs a useful cognitive model of how critics produce successful interpretations.[15] Critics here are seen as craft practitioners, and the construction of meaning in interpretations is "a routine institutional activity, a body of ongoing craft practices that draws upon abstract doctrines in an ad hoc, utilitarian, and 'opportunistic' fashion" (Bordwell, 1989, 27). This characterization is meant less to imply that the critic is cynical in producing the type of interpretations demanded by the institution of academic criticism than to imply that one tends to approach the task with a tacit understanding of the norms of novelty and persuasiveness that prevail in the institution, hence with a set of more or less well defined goals in mind. Within the prevailing limits of plausibility, it is perfectly acceptable for the critic to subordinate an abstract doctrine or theory to the requirement to reach a goal congenial to the institution. Theories, like psychoanalysis or the topological concept of neighborhood serve as a source of the highly flexible schemata that mediate in a practical process of negotiating between literary texts (such as Maihime and *Madame Bovary*) and institutionally sanctioned semantic fields (such as the opposition "Japan/the West").[16]

Had Maeda allowed the interpretation to be guided by his own critical construct, he should have been led to gender as the relevant difference. However, he uses theory in an ad hoc manner to produce an interpretation in terms of national culture. In this understanding of the critic as craft practitioner, to say that topological analysis has not been applied in a conceptually rigorous way does not mean it has not been a useful and effective tool for the routine institutional task of constructing novel and persuasive interpretations. The essentially practical nature of interpretation stresses success in mapping semantic fields onto details of the text over questions of conceptual

15. Although Bordwell's argument is made in reference to film criticism, he claims that the interpretive skills he describes "are not, of course, specific to film criticism. They derive from twentieth-century criticism generally, and especially literary interpretation" (1989, 32).

16. Sakai (1990), 64-73, discusses the tendency in Japanese scholarship to accept theoretically unexamined oppositions between the putative unities of "The West" and "Japan." See especially p. 72.

rigor the notion of "theory" might suggest.[17] Hence, since Toyotaro and Mitsugi, while clearly moving across borders, do not seem to define them, the neighborhood schemata featured so prominently in the discussion of *Madame Bovary* may be dropped in discussing Maihime and *Teriha Kyōgen*. In section three there are new goals, and neighborhood is reintroduced and manipulated to suit the task at hand, producing a compelling critique of Lotman's binary reasoning by emphasizing the undecidability of the boundaries neglected in section two.

The citation of topology also provides a feeling of scientificity, masking the loose, malleable, and goal-oriented nature of literary interpretation by a rhetoric of disinterestedness and precision. For example, when setting out the relation of topology to literature, Maeda refers to it as a "hypothesis" involving the selection of "a random sample from a continuous act of reading." This argument suggests that *Madame Bovary*, far from being a random sample, was selected especially for the purpose, and that this is typical for professional literary criticism.

This brings the argument full-circle to the question posed in the opening of this paper, of what is at stake in the superficial citation of topology in Japanese literary criticism. Van Rees claims that the unargued citation of specialized terminology often points back to a more thoroughly argued instantiation of the terms, with the important qualification that subsequent critics will be inclined to adopt not only the proposed terminology, "but also the premises implicit in these terms." One must raise the possibility that the intrusion into Maeda's argument of easy oppositions between Japan and the West marks the set of premises called up by the fashionable citation of topology today, that is to say, the easy exit from modernity. The citation of topology today does indeed gesture back to Maeda's problematization of traditional logic and Euclidean space, however shorn of Maeda's theoretical struggle, what resonates is institutional common sense.

---

17. Even in the sciences, the relation of theory to the object of study is an open question, and the projective role of the investigator enters the problem in both the early and normal stages of science. Kuhn argues, for example, that normal science is, above all, a puzzle-solving activity, and that "no process yet disclosed by the historical study of scientific development at all resembles the methodological stereotype of falsification by direct comparison with nature." See Kuhn (1970), 77-78. This is not the same, however, as saying that the knowledge produced by science has an arbitrary relation to the object.

The Work of Theory

> [I]n general geometry is the place where concepts from the most
> different regions get mixed up together.
> —Wittgenstein, *Philosophical Grammar*[18]

If the failure to follow through on the critical schema actually turns out to be a success, the theoretical problem is more fundamental. In that the clarity of Maeda's exposition bares a certain device, it is useful to work through the problem. What Maeda has proposed to do in bringing together topology and literature is to borrow a set of concepts developed in one domain, mathematics, to consider the elements of another domain, literature. Specifically, the categories and operations of a topological analysis developed in the domain of a metric space based on the continuity of real numbers are to be brought to bear on the "textual space" of the literary work of art. Geometries have always been a privileged ground for this kind of mixing of theories, enabling the translation of a multiplicity of domains of experience into the palpable, manipulable and easily visualized relations of space. When successful, this often leads to a proliferation of knowledge in the field to which it has been applied.

What is crucial for a juxtaposition of disciplines to function in this way as a switching yard, where methods and procedures of one theory can be used to manipulate and consider elements from another system, is the establishment of a one-to-one correspondence between domains (Matteuzzi, 1995; Berlinski, 1995). Hence the ur-problem in putting calculus on a sure footing as a tool for mathematical expression of relations in the physical world was the reconciliation of the discontinuity of the domain of integers used in algebra (. . . 1, 2, 3, 4, 5 . . .) with the continuous nature of space and motion in experience. Until the continuity of the *domain* of mathematical notation could be established, that is to say the continuity of real numbers, any attempt at a mathematical expression of the continuous relations of space and motion would end like Zeno's paradox in contradiction. The huge gaps between integers were quickly filled with what were called the rational numbers (that is fractions, or *ratios* of integers), however there still remained certain tiny but persistent gaps and inconsistencies at, for example, zero, π or the square root of 2. This was a source of considerable anxiety until the solution of irrational numbers by Dede-

---

18. Kenny, ed., *The Wittgenstein Reader* (1994), 40.

kind in the nineteenth century. This produced the continuity of *real* numbers, authorizing the mathematical expression of nature on which the adventure of the scientific revolution had been proceeding on faith since Newton.

This production of hybrid theories through the mapping of domains and operations is an absolutely fundamental process in the construction of knowledge. Building a successful correspondence between different conceptual systems tends to result in the extension of both disciplines concerned into new territory, as the corresponding operations push and pull each other into new associations and new conceptual structures. One need only recall the project of semiotics to see its importance for the humanities. Efforts to develop, for example, a linguistic model of film begin with Christian Metz in *The Grand Syntagmatique* where mapping the linear and discrete succession of frames in film projection to the linear succession of words in language enabled systematic analysis of film "language" through the operations of syntagmatic and paradigmatic substitution in structuralist linguistics. Such "exchange" enables one to borrow the systematicity of linguistics to produce new significance and insights, and extend analyses in unforeseen directions (if film is indeed structured like a language, for example, what corresponds to the "subject of enunciation?" See Bordwell 1989).

In a discussion of complex numbers (which combine real and *imaginary* numbers) in the context of themes of cultural multiplicity and association, Andrew Pickering again identifies the crucial move in allowing a spatial intuition of the absurd quantities:

> [T]he move to construct an association between algebra and an otherwise disparate branch of mathematics—geometry . . . consisted in establishing a *one-to-one correspondence* between the elements and operations of complex algebra and a particular geometrical system. (Pickering, 1995, 423-424, italics in original)

One need only substitute topology and literature for algebra and geometry to see Maeda's project. The intuition that sets this project in motion can be discerned early in the theoretical introduction to *Literature in the City Space*.

> The space traversed by the reader is, like space in dreams, contracted in parts and expanded in others. The space that is closing in and oppressing the reader's heart in one place will burst wide

in freedom in the next instant. This is the kind of open space the literary work underwrites, and it is far removed from our everyday world. (Maeda 5)

Maeda can be seen here gesturing to the spatial malleability that topology captures in a way that suggestively echoes the efforts of Michel Serres to bring topology into the study of history. The whole project, with its insights, its flaws, its important problematization of Cartesian space and time, is contained in this sentence. Lurking behind the statement, however, is the highly questionable assumption that continuity, as an attribute of experiential space, will also pertain to the represented space of literature. Maeda makes this explicit several pages later: "If one takes a seamless continuity to be the most essential characteristic of space, then this characteristic will apply to the "interior space" of the text as well" (11).

Maeda maintains that, though language introduces the space of the text through a succession of discrete specifications (the wall, the table, the cup, Emma's shoulder), continuity in the represented space of the text is restored through the act of reading. This would seem to be a fundamental error from the perspective of Ingarden's phenomenology of literary experience. The imaginational space constructed through the act of reading is essentially different from lived space in its concrete development in that it is structured with gaps and "spots of indeterminacy." According to Ingarden, these are, "entirely separate kinds of space, between which there is no spatial crossing."[19] This is something Japanese critics, trained to approach literary criticism as part of the larger project of knowledge, spotted right away. Here is editor Taki Koji, from the critical afterward to Maeda's collected works:

> The theoretical approach enabled by the neighborhood concept is extremely original. However, one can discern there the premise that the literary text is a topological space. (Taki, in *Kaisetsu* to Maeda 1989, 434-435)

This disparity between the continuity of the domain of topology and the discontinuity of the domain of literary space calls into question the entire project. Maeda dispenses with this crucial point by fiat, and the repeated return in his argument to an ideal scene of reading where one's absorption in a book is so intense that all consciousness

---

19. Naoki Sakai drew my attention to the fundamental nature of the problem of discontinuity in Maeda's work. For the argument on "Spots of Indeterminacy of Represented Objectivities," see Ingarden (1978), 246-255.

of the world around is lost, where in effect the story world comes to life, appears as a symptomatic attempt to cover this discrepancy through the Herculean efforts of the individual reader.

Geometry can be described as the place where concepts from "the most different regions" get mixed up together. The juxtaposition in Maeda's argument of the purest expressions respectively of science and the humanities would appear to be pushing the notion of cross-disciplinarity past its limits, I would argue that there is nothing essentially unsound about the project of bringing topology and literature together. It is in the requirement for a *one-to-one* mapping between disparate domains, not the degree of disparity itself, that Maeda's effort runs aground in contradiction. The repressed problem of discontinuity returns in the mapping process itself. While generating the hybrid object of theory requires mapping a correspondence from domain to domain, what Maeda is actually doing is mapping a correspondence from domain to theory, from the necessarily schematic axioms and operations of topology such as neighborhood, onto the schematicity of representational space.

Conclusion

Paraphrasing Fredric Jameson, we may say that representations of theory, like representations of death, are always ultimately displaced meditations on something else. This chapter was the first written in this book, and finds its displaced object in the disposition to buzzwords in critical discourse. The initial conjecture was that the fashionable citation of "topology" as a critical term in Japanese criticism would have behind it a serious engagement, and that this engagement would be shaped by the provision of a certain math-literacy in the Japanese humanities by the rigors of the examination system. One certainly sees in the introduction to *Literature in the City Space* a willingness to present a mathematical subject without caricature or reduction to an authorizing metaphor. However, what becomes salient on closer examination is rather a tendency for institutional demands to override the function of theory in purportedly critical work. Subsequent fashionable references to topology in the 1990s, shorn of Maeda's struggle with the theory, are pure suture into the fabric of institutional common sense, raising vague anxieties about the reader's competence, then offering to make good that lack through identification with the community of theoretical cognoscenti.[20] By pressing on a seemingly in-

---

20. Suture is "the process whereby the inadequacy of the subject's position is

nocuous bit of jargon I have suggested that behind its ring of authority is a kind of call to conformity in the institution of literary criticism that nods its head at any string of informal reasoning as long as it ends up in an institutionally sanctioned semantic field.

This paper is secondly an effort to think through the role of "theory" itself in the discourse of criticism. Theory, by which I mean systematic conceptual structures imported from fields in which they have been built up from first principles, can potentially do for literary studies exactly what it does for the sciences, that is, "provide a guide . . . to take them into unfamiliar territory" (Feynman 1995, xii).[21] There is no reason why math and science, articulated with imagination and rigor, cannot perform for literary studies this same function of leading one's speculation someplace unexpected. Conversely, however radical the politics a theory thematizes, its routinization within an institution spells the end of its theoretical function.

Given this fundamental affinity in the way theory is deployed in the humanities and in the sciences, there is a further point of similarity and a point of difference. On the one hand, it would be difficult to say that the intrusion of institutional factors into the production of a putatively "objective" knowledge is a problem unique to literary studies. Whenever a new theory begins gaining ground in the sciences, one finds a constant questioning of whether its introduction is determined by genuine scientific issues (economy, predictive value, ability to suggest new lines of research), or whether it is responding rather to fashion, the scramble for grant money, or the demand for novelty in publication. The semantics of computer programming languages, for example, were shown to be amenable to topological treatment as early as 1971. This was criticized, though, as "merely a technical trick," a dubious status it held for some 10 years, until topology, newly expressed as a logic of semi-decidable properties, was shown to "capture[s] an essential computational notion" (Vickers 1989, 1-13). Similarly, the respondents to an influential early paper claiming chaos as a functional process in sense recognition consistently hammer at the status of chaos in the analysis with reference to notions such as metaphor, fashion, and "fad or insight" (Skarda and Freeman, 1986,

---

exposed in order to facilitate (i.e., create the desire for) new insertions into a cultural discourse which promises to make good that lack" (Silverman 1986, 234).

21. Cf Anatol Rapaport: "Game theory, we think, is useful in the same sense that any sophisticated theory is useful, namely as a generator of ideas." Quoted in Lyotard (1983), 60.

173-192, especially responses by John A. Barnden, Robert Brown, Donald H. Perkel and René Thom).

Where science radically differs from the humanities, though, is in the notion of what constitutes knowledge thus produced. In Feynman's operational definition, the defining characteristic of science is that "the test of all knowledge is experiment." All kinds of speculation is produced when following out the implications of a theory. But if an experiment cannot be devised to make nature give back the same pattern of events, they will fall by the wayside. Lyotard maintains this difference while making the apparently postmodern gesture to collapse the difference between literature and science in the aforementioned, "Postmodern Science as the Search for Instabilities," qualifying the characterization of the scientist as above all "a person who tells stories," by adding a difference: "The only difference is that [the scientist] is duty bound to verify them" (Feynman 1995, 2; Lyotard 1983, 60). But that is a significant difference, and there is no corresponding notion in literary studies, hence the ease with which institutional norms overpower Maeda's ostensibly rigorous analysis.

Yet, something is at work, in the institution, checking the results. Maeda's work did, like bad science, fall by the wayside, tactfully displaced by the editors of the posthumous *Maeda Ai Chosakushū* (Collected Works, 1989) from the solid and productive insights of his *Literature in the City Space* to the more speculative sixth volume. Yet the author of the critical afterward to this volume, who clearly recognizes the problem that lies in the mapping of topological space and literary space, stresses over and over, as compensation for the theoretical shortcomings Maeda's skill and sensitivity as a reader. My argument is that it is precisely this institutionally structured "critical common sense" that is in some senses inimical to the use of theory.

But though it can be read as a displaced meditation on other things, this paper is also really about topology, and about the possibilities of a communication between science and the humanities in a time when a trend toward a "complex" view of knowledge in both fields seems intuitively to be bringing them onto common ground. This reading never forecloses the possibility that someone will construct a bridge in the future, and in fact, by its very failure Maeda's attempt to engage topology shows us the domain to domain mapping by which this might be done. It seems to me, though, that such an approach, if indeed possible, needs to think through the question of discontinuity, and will need either to link up with the work of Michel Serres in the philosophy of history, or use the notion under which computer science has incorporated topology, that of open sets as semi-decidable properties (Vickers, 1989, 1-11). This need not be restricted to spaces that can be "visualized" like the represented space of the text, and might better be applied to the actions of the reader or critic as a reader response technique, studying the process of "deciding" the undecidable text as a "logic of finite observations."

6 Contemporary Stakes: The Science Wars

Any statement about the relation of literature and science in the North American academy must seek its stakes in the barren terrain known as the "Science Wars." Though it seems expedient to write this embarrassing dispute off as a media spectacle, this chapter seeks to discern a longer history to the Science Wars, which casts it neither as an accident, nor a failure on the part of its participants, to be corrected with further application of academic protocols. The repetition of episodes like this rather discloses the same gap in the field of knowledge toward which each of the thinkers in this study were driven, and on which, as Ian Hacking has argued, reasonable people may be forever fated to find no common ground. It pays to take the time to understand the relatively productive engagement between literature and science visible in twentieth-century Japanese writers, but this is not a hopeful study.

The Science Wars names both a specific media event in the late 1990s, and a general impasse that has arisen between two cultures in the contemporary university, the sciences and a theoretically oriented humanities. C. P. Snow somewhat tendentiously mapped out this terrain for the Cold War West in the Two Cultures debate in 1959, positing "[l]iterary intellectuals at one pole—at the other scientists, and as the most representative, the physical scientists." Snow's faith in progress, the ability of instrumental science to solve problems of unequal distribution being revealed as systemic in the postcolonial world, and explicit call for a proselytizing mission by the western intellectual center mark his uncritical allegiance to then-current modernization theory, while the vitriolic reply by F. R. Leavis stands as a near-hysterical symptom of discredited notions of the Western high tradition as monument to the "fully human mind" (Snow 1959, 4; Leavis 1972). In the same special issue of the *Stanford Humanities Review* from which Norman Holland diagnoses the reluctance of humanists to engage scientific approaches to literature cited in chapter 1, Brian Rotman expressed the belief that surely the simplifications represented in Snow's analysis of the nexus of technology, science and culture, and the impasse evident in Leavis' reply had long been overcome (Rotman 1994, 99). But despite optimistic talk of a "Third Culture," and despite claims that an interdisciplinary field of literature and science had become "an established point of departure" in the university, it would be that same year that the science wars would emerge with the publication by Gross and Levitt of *Higher Superstition*, a sustained attack by a mathematician and a research biologist on the competence of the study of science by the theoretically informed humanities (Gross and Levitt 1994; Shaffer 1998, 1). The science wars is an extension of the basic Two Cultures debate, where what is now known as a theoretical approach in the humanities stands in for Snow's broad term "literary intellectuals."[1] The social sciences occupy an ambiguous position in the science wars.

The Two Cultures debate in 1959, and the science wars in 1995 are elaborations of the same problem, a split in the modern university between the sciences and the humanities as means of producing knowledge. What distinguishes the science wars from the more personal polemic of the Two Cultures debate is a proximate cause that intervened in the 1970s and 1980s in the form of a move by the humanities to claim the discourse of science as a subset of its own concerns. This move begins with the diffusion of what is sometimes called a "constructivist" or "constructionist" position in the critique of science. Researchers in the history and sociology of science have been probing the idea that the facts produced in scientific research have an irreduc-

---

1. "Literature" for Snow, like "the academic left" for Gross and Levitt, is understood to stand in for the culture of the humanities. "[L]iterary intellectuals, who incidentally while no one was looking took to referring to themselves as 'intellectuals' as though there were no others" (Snow, 4). Both "literature" and "philosophy," while naming specific disciplines in the modern university carry with them an older history where they stand for letters or learning in general. I will typically use "literature" or "the humanities" interchangeably, with the understanding that the opposition "literature and science" only carries meaning when the modern discipline seeks, as Snow intuits, to reclaim its old generality.

ibly social and political component at least since Thomas Kuhn's *The Structure of Scientific Revolutions* troubled rational-realist models of scientific knowledge in the early 1960s, though the roots of this position can be traced within science to the writings of Mach, Duhem, Poincaré and Bohr in the early twentieth century. The constructionist legacy begins with the inheritance of Kuhn in the so-called strong programme of the Edinburgh school, and crossing over to France and the United States in the 1970s, proliferates into a variety of disciplinary method-ologies and concerns, incorporating on the one hand ethnographic, sociological, and rhetorical analyses of scientific practice and on the other feminist and culturally relativist critiques of the forms of scientific authority. What perhaps unites these disparate efforts is the deploy-ment of an anti-foundationalist critique against the claims to privilege and authority of scientific descriptions of the world.[2]

Over the same period, there emerged over a spectrum of disciplines in the humanities, but centered in departments of literature and an-thropology a tendency toward a political problematization of univer-sality, objective knowledge and canons of value identifiable in retro-spect as the rise of theory. Theory is understood here not as a poetics or theory of literature, but in the way it has been institutionalized in the humanities since the 1970s as what may fairly be called a kind of "superdiscipline about human consciousness and culture" (Rickman 1996, 57-58).[3] The constructionist position in science studies, then, provided a context where a theoretically-informed cultural criticism in the humanities was eventually bound to turn its gaze toward the natural sciences as the site where the problematization of universality and objective knowledge implicit in the project of theory would have the greatest polemical effect. Proceeding as a kind of enthusiasm from the bracketing of the question of reference in structuralism, and the elevation in deconstruction of the literary as a model for a general textual indeterminacy, a tendency became visible over the 1980s in humanities-based discussions of science, often labeled poststructural-ist, postmodernist or the cultural studies of science, to understand the constructionist critique to deny any difference whatsoever to the discourse of science.

---

2. This movement is usefully summarized in Golinski (1998), while Hacking (1999) delineates the critical stakes of the anti-foundationalist position. The set of positions is usually called constructivist, but I follow Hacking in using constructionist.

3. On the transformation of other humanist disciplines by the terms, methods and assumptions of literary theory See Simpson (1996).

This can be tracked in two ways: in the rhetorical figuring of science as a narrative activity, and in efforts to locate metaphor in the processes of scientific thought. Taking first the former strategy, in an influential primer on the problem of postmodernism, J. F. Lyotard characterizes the scientist as before anything else "a person who 'tells stories'." Anthropologist Sarah Franklin sees the rise of "poststructuralist, deconstructionist, psychoanalytic, and postmodern theory" placing special tools in the hands of the cultural critic adequate to discerning the significance of science writing, and literary critic Terry Eagleton can speak the common-sense of the field without argument when he says that science must learn to view itself "more modestly as just another set of narratives."[4] Franklin makes explicit the new position humanities researchers imagined prepared for themselves: "Long enshrined as a kind of apex of rational knowledge production . . . science is now up for deconstruction just like all the rest of the Western canonical fare" (Harvey 1990, 9; Franklin 1996, 142; Lyotard 1983, 60).

Let us examine this claim. It is neither surprising nor pointless for literary criticism to look at the verbally elaborated field of science and see narrative, metaphor, and the shaping of these expressions by power relations. The problem for anyone interested in the relation of literature and science lies in the leap to "just another." The notion that scientific facts are socially constructed does not of course mean that they are fiction, nor does it follow that the reconfiguration of the old terms text, discourse and narrative to a more general model of signifying activity in thinkers like Lyotard, Foucault, and Derrida, reduces them to an object of the academic discipline of literary criticism, or even a hermeneutics broadly conceived. There is further the problem of descriptive adequacy. Such reductions offer no way of accounting

---

4. "Science and philosophy must jettison their grandiose metaphysical claims and view themselves more modestly as just another set of narratives." Quoted in Harvey (1990). Though the Marxist Eagleton undoubtedly intends some irony, he is quoted with full approval by Harvey, who lists as symptoms of the death of metanarrative shifts in the philosophy of science "wrought by Kuhn (1962) and Feyerabend (1975)," and "recent developments in mathematics emphasizing indeterminacy "such as chaos theory and fractals, unfortunately both compatible with determinacy. In "Scrutinizing Science Studies," Noretta Koertge gives a fair articulation of this position, which for her the opposite camp: "Although scientists typically succeed in arrogating special epistemic authority to themselves, scientific knowledge is just 'one story among many.' The more epistemological authority that science has in a given society, the more important it is to unmask its pretensions to be an enterprise dedicated to the pursuit of objective knowledge. Science must be 'humbled'" (Koertge 1998, 3).

for differences between science and other kinds of discourse, in particular its instrumentality, ability to increase technical performance over time, and consequent value to institutional sources of funding, as evidenced by the fact that government grants for the sciences and engineering are on the order of 180 times the value of grants in the humanities.[5] There is reason to accept the boast by one commentator that from a conceptual or institutional point of view, "the sciences are in an unprecedentedly robust state of health, strength and vigor" (Levitt 1998, 280). The mainstream scientist has an empiricist dissatisfaction with the inability of the constructionist project to account for the grotesque dissymmetry between the humanities and science: "To the extent that one denies objectivity to science . . . one is forced to find more sinister explanations based on the foibles of individuals."[6] The move to objectivity needs to be scrutinized, however the author is correct in implying that any theory that requires the magic wand of false consciousness to account for a dissymmetry of this order of magnitude is inadequate as an analysis of the politics of knowledge in modern society. Scientists simply do not take seriously the claim that scientific facts are persuasive because they tell a good story.

Lyotard himself, whose statements about postmodernism have been relentlessly caricatured in the science wars, ends his discussion of the relation between complexity theories and performance enhancement in *The Postmodern Condition*, from which the previous quotation is taken, with an enigmatic reservation of a difference for science. The scientist can indeed be conceived as a person who tells stories, however "the only difference is that he is duty bound to verify them" (60). This qualification of a difference, and the calculated ambiguity

---

5. James Herbert, director of research and education for the NEH puts the figure at 33 times the dollar amount, but only considers the National Science Foundation (Paulette W. Campbell, "NEH Official Calls for Broader Research into U. S. Humanities Policy," *The Chronicle of Higher Education*, June 25 (1999), A42-A43). According to a budget report in *Science* (Mervis, Jeffrey, "2001 Budget: Spending Bills Show No Sign of Surplus—Yet," *Science* 289, no. 7 July (2000), 31), the combined research outlay for FY 2001 for the National Institute of Health (NIH), National Science Foundation (NSF), NASA, the Departments of Energy and Defense, and the National Institute of Standards and Technology (NIST) is projected to be on the order of $54 billion. Compare the 2001 Budget request for the NEH and the NEA at $150 million each (sources: NEH website: <http://www.neh.fed.us> search "2001 Budget"; NEA website <http://arts.endow.gov> 2001 Performance Projection). This figure has sextupled in the last 20 years, and does not include corporate funding.

6. From an unfavorable review of Mara Beller's iconoclastic, constructionist reading of Bohr and Heisenberg in *Quantum Dialog*. See Greenberger (2000), 2166-2167.

of the term verify closely recall Richard Feynman's operational characterization of science: "[T]he principle of science, the definition almost is the following: The test of all knowledge is experiment" (Feynman 1995, 2). That is to say, science, like the humanities, produces speculation. However, if it cannot be tested by experiment, it does not count as scientific knowledge. This reservation of a broadly specified but intractable difference for science, and the care with which Lyotard delineates the natural and human sciences were lost in the hurried translation of his advocacy of chaos, catastrophe and fractals to the North American marketplace in the 1980s. By the 1990s, a resistance on the part of natural scientists to this invitation to keep company with literary criticism was clear.

The Ground in Literary Criticism

This conceit by a theoretically informed humanities to have annexed science as a subset of its own concerns has a history within the United States that involves a caricature of the scientist as a naive positivist.[7] The notion that science is the province of unambiguous statements about an objective world goes back to the repositioning of the irrational, the shadowy and the unconscious in the romantic period from a marginal to a central component in the definition of literature. With the advent of the New Critics in the mid-twentieth century, this self-definition took its most clear form as a kind of gentleman's agreement within the university to coexist. According to Gerald Graff:

> The influential group that came to be called "New Critics" in America and England after World War II argued that whereas ambiguity may be a fatal defect in a laboratory report or accounting ledger, it is a necessary and valuable attribute in a literary work.

---

7. This in itself is only intelligible within a longer history of opposition in the university between positivist and hermeneutic approaches. For an account of the movement of late nineteenth and twentieth-century thought as "a kind of historical see-saw, an oscillation to-and-fro" between variants of a basic hermeneutic and positivist position, see Bhaskar (1979), 17-19. Bhaskar's description of waves of action and reaction in the twentieth-century set in motion by the reaction of Wittgenstein, Russell and Moore against the idealism of the intellectual milieu of Dilthey makes the notion that appears in the 1970s and 1980s, that science could properly be an object of literary criticism, immediately intelligible as a play to extend activities once used to contrast the social sciences from the physical sciences (the elucidation of meanings and tracing of conceptual connections) to the center of the natural sciences too, making the entire field of human knowledge an object of literary criticism. Absurd on the face of it, it is a predictable institutional move.

Whereas science speaks directly by means of propositional statements that aspire to have one and only one meaning, poetry
speaks through metaphors or images, which multiply the meanings rather than restrict it. (Graff 1995, 164)

Here the poet or critic and the scientist acknowledge their differences and agree to coexist. What has changed since the time of the
New Critics is that literature is no longer content with separate domains, but wants to come over and inform the unimaginative func-
tionaries of science (and Graff's lining up of science and accounting
is not accidental here), that if they look they'll see that their project is
underlain by the same ambiguity and indeterminacy.

Concentrating just on the question of metaphor raised by Graff,
it is easy to show the way this proceeds by a caricature of the scientist
as a nineteenth-century positivist. Central to this claim is the idea
that scientists see metaphor as a kind of soft, degraded use of language
inimical to the process of science. For example, in an account of the
history of the constructionist critique of the sciences, Peter Golinski
writes, "Scientific language works to persuade its audiences that they
can read *through it* to apprehend nature," (Golinski 1998, 104, italics
in original) while in a volume called *Literature and Science: Theory and
Practice*, N. Katherine Hayles writes that from the scientist's viewpoint
"metaphors have not been admitted as components of the scientific
process" (Hayles 1990, 211). And again in a discussion of the popularity
of chaos theory across a range of audiences from lay reader to scientist
Weingart and Maasen write:

Only one domain of communication is—in theory—excluded from
the occurrence of metaphors: scientific communication. Being
vague, polysemous constructs, metaphors deeply insult the positivist's ideal of objectivity. (Weingart and Maasen 1997, 475)

The locus classicus of this notion lies in statements by the Royal
Society in the 1660s issuing to members a prohibition against using
metaphor in science writing. And this point of practice no doubt
played a key role in the generation of the uncertain identity of modern
science through the delimitation of classical notions of rhetoric. But it
is just as easy to show that, whatever the popular conception of
science may be, and whatever programmatic statements can be located
in the 17th century, working scientists today are quite explicit in
acknowledging the role of metaphor in their thought, and comfortable
talking about it. This from a discussion of chaos as a strategy for
modeling the sensory activity of the olfactory bulb:

The results indicate the existence of sensory and motor-specific information in the spatial dimension of EEG activity and call for new physiological metaphors and techniques of analysis. (Skarda and Freeman 1987, 161)

In "The Pool-Table Analogy with Axion Physics," physicist Pierre Sikivie uses metaphor as a tool to construct a thought-experiment about the existence of the axion, a tenuous particle for which there is no direct experimental evidence. "Why should one believe in the axion? I attempt to answer this question by drawing an analogy with the physics of a pool table" (Sikivie 1996, 22). And the special issue of *Science* announcing the sequencing of the human genome by Venter et al. includes a section on "Metaphors and Meanings."[8] Moving from pure science to engineering, in explaining their inclusion of a lengthy digression on the work of Heidegger, Gadamer, Maturana and J. L. Austin in a book on artificial intelligence design for engineers, Winograd and Flores write:

Readers with a background in science and technology may find it implausible that philosophical considerations have practical relevance for their work. Philosophy may be an amusing diversion, but it seems that the theories relevant to technological development are those of the hard sciences and engineering. We have found quite the opposite. Theories about the nature of biological existence, about language, and about the nature of human action have a profound influence on the shape of what we build and how we use it. (Winograd and Flores 1985, xii)

I cannot see how these two facts; the pervasive caricature of the scientist as nineteenth-century positivist that enables the humanities-based critique of science, and the sophisticated attention of working scientists to questions of metaphor, language, and the relation of conceptual systems to notions of objectivity and the physical world can be reconciled. This straw target circulates unopposed in

---

8. "The sequencing of the human genome affects how we think about ourselves. . . . Recent books have already begun to chronicle the impact of genome sequencing, and three have been reviewed in this issue. Kay has examined how the metaphors of information, language, and code influenced the research program and claims of molecular biology. In his book review, Lewontin (p. 1263) believes biologists should not be dissuaded by her poststructuralist terminology because she effectively demonstrates why the metaphors can be misleading" (Jasny and Szuromi 2001, 1155-1157).

humanities-based discussions of science, and the perception that the
Japanese literary critics I encountered were not nearly so quick to
reduce science to a tractable caricature was the starting point of my
thinking on this subject.

The result of this history is that what appears to be the interdisciplinary undertaking par excellence: exploring the relation between
literature and science, is effectively carried out in the absence of the
other discipline. Scientists and engineers appear in the humanities-
based critique of science not as colleagues, but as straw targets, and
the discourse engages not science writing, that is to say technical
papers as circulated in specialized communities, but popularizations
and broadly drawn metaphors, planed of their subtleties and circulating far from the communities that produced them.

Visible in statements by working scientists and engineers is a
reflection on the relation between the social question of what is guiding
scientific and technological research and the epistemological question
of the status of the knowledge produced. The finessing of this distinction in science studies is a virtual constant of the genre. In a discussion
of the formative influence of Kuhn's *The Structure of Scientific Revolutions*, Evelyn Fox Keller writes:

> More than any other work, it was here that the intellectual space
> was cleared for responsible scholars to examine the formative
> role for scientific inquiry of its particular social and political con-
> texts. *Structure* provides a launching pad for social studies of
> science. . . . If scientific knowledge was dependent on social and
> political forces to give it direction, *and even meaning*, then it was
> surely reasonable to suppose that 'gender,' which exerts so pow-
> erful a force in shaping other parts of lives and worlds, would
> exert its force here as well. (Fox-Keller 1998, 17-18, italics added)

The slip in the dependent clause, from the question of what gives
science a particular direction and conditions the imaginative act of
producing hypotheses to the much more difficult question of the
status of the knowledge produced is a virtual constant in this genre,
and left unargued. And it is this slip that allows the movement from
the social scientific investigation of science as practice, the pertinence
of which few scientists or engineers would seriously question, to
more dramatic claims of the deconstruction, narrativity or inherent
metaphoricity of scientific knowledge itself. Analytically separating
the question of what guides scientists to investigate this problem,
instead of that one, from the question of the nature of the knowledge

thus produced doesn't render scientific knowledge free from contamination by the social. But the constant slippage in strong constructionist literature between points made about the former to claims made about the latter vitiates the position for all but the already convinced.

These, then, are the conditions I see preparing the ground for that outcropping of the academic unconscious in the mid-1990s known as the science wars. Literary criticism over the last 25 years has refashioned itself under the rubric of literary theory from the relatively well-delineated study of poetry and fictional texts of a certain quality, into a kind of superdiscipline about human consciousness and society. In this model, arguably traceable to dated but consistent notions of the determination of thought by language,[9] whatever the philosophical sophistication of the notion of textuality in theory, all other disciplines become subject in practice to the same semiotic and formalist analyses as literary texts, which expose their presuppositions and demonstrate that their protocols lead ultimately not to sure, foundational knowledge, but to the same indeterminacy as literary texts. Under such a program, readable as an effort to regain a generality enjoyed by literature prior to its specialization in the modern university without asking into the historical change in competencies such a division of labor instituted, humanities-based cultural critique has sought to comprehend science in the sense of subsuming as part of a larger whole without understanding science, to wit, without engaging the question of mathematical representation. Whether this is ascribed to the belief that skeptically exposing the epistemological foundations of science

---

9. The thesis of the identity of thought and language is a hypothesis about meaning which argues that, though one may have a subjective sense of the richness of radically private experience, because one cannot observe the contents of another person's mind it is words, and not concepts or ideas, which must form the basis of any statement about thought, because that is what is out there and available to intersubjective inquiry. The thesis first appears in John Horne Tooke in the late 18th century, and in the early Schleiermacher for the hermeneutic tradition, and arguably has provided for 200 years the most consistent underpinning to the various forms of the insistence in contemporary theory that it is "language, all the way down." With the development over the last 20 years of non-intrusive methods of observing the brain at work, however, such as PET scans, MRI's, etc., contemporary cognitive science, in a break with behaviorism, has begun to forcefully advance the position that there is a "mentalese" different from articulation in language, a position which benefits from congeniality with subjective experience. Pinker's *The Language Instinct* (1994), 44-73 carries on a sustained, low-level polemic against the linguistic determination thesis in the humanities. See Quine, *From Stimulus to Science* (1995), 4-7 on Hooke's reformulation of Locke; and Palmer (1969), 84-97 on early Schleiermacher.

allows its content to be bracketed, or whether it is seen as a practical evaluation of the strategies open to a humanities that has systematically eliminated math literacy from its repertoire, it has meant in practice that the humanities-based critique of science tended to proceed by engaging straw targets of its own making, or popularized or dated historical texts, rather than actual, contemporary science writing. The result is that the critique is not taken seriously outside the humanities.

Stage One: A Valid Point

The science wars begins in 1994 with the publication by Gross and Levitt of *Higher Superstition: The Academic Left and Its Quarrels with Science*, a really damaging attack by a mathematician and research biologist on the variety of critiques of science that had been developing without serious opposition in departments of English and Comparative Literature as part of the rise of poststructuralist theory from the 1970s. What makes the attack so damaging, and what tends to drop out in the traumatized memory of the encounter, is the remarkably fair engagement of the concerns of contemporary theoretical critiques of science by which they precede each discussion. If you want to understand your field, look at the summary provided by Gross and Levitt in *Higher Superstition*. However, in a series of detailed analyses of work published by major university presses, Gross and Levitt make the case that forays into the discussion of science and technology by what they call postmodern critics fall in a range from superficial to incompetent, and more damaging, have nothing of interest to say to scientists.

It was after reading *Higher Superstition* that New York University physicist Alan Sokal conceived the idea of submitting a faux-theoretical discourse on "the historicity of π" to a leading journal in the cultural studies of science called *Social Text*. The subsequent publication of Sokal's clumsy parody of science studies cliches as the capstone article of a special issue on "The Science Wars," revelation of the hoax by Sokal and ensuing media scandal undermined at birth the response by humanists to *Higher Superstition*.[10] George Levine, professor of

---

10. Sokal, Alan D., "Transgressing the Boundaries: Toward a Transformative Hermeneutics of Quantum Gravity," *Social Text* 45/46, Special Issue on The Science Wars (1996), 217-252. The fifteen other contributions to this special issue, though tainted by association, remain worth reading for the array of genuine responses to the challenge of *Higher Superstition*. The revelation of the hoax appears in Sokal (1996b), 62-64. *Lingua Franca* ran a series of responses to the incident by leading figures in science studies under the title "Mystery Science Theater," *Lingua Franca* July/Aug (1996),

English and Director of the Rutgers Center for the Critical Analysis of Contemporary Culture put the case with unintentional irony in the same issue of *Social Text* in which Sokal's parody ran: "We are, self-evidently, at a moment of aggressive public attack . . . on science studies." The humanities has never managed a coherent response to the problems raised here, and the most substantive reply remains Evelyn Fox Keller's admission in the issue of *Lingua Franca* following the Sokal hoax that no amount of misgivings about the strategy or motives of the scientists involved "condones the failure of my colleagues in science studies to acknowledge so blatant a compromise to the integrity of their own discipline" (Fox Keller 1996, 58; Levine 1996, 113). Though there are many ways to dismiss the intellectual significance of this event, I would argue this is a fiasco the humanities dismisses at the peril of its own increasing marginalization in American discourse.[11]

This brings up an ambiguity. The field of "science studies" or "science and technology studies" stretches across the humanities and social sciences and encompasses a variety of disciplines, including history of science and technology, literary and cultural criticism, women's studies, cultural studies, anthropology and sociology of science. What unites these disparate strategies and intellectual affiliations—versus the broadly foundationalist conceptions of traditional philosophy of science which would explain the coherence of science as a collection of ideas by reference to a world existing independently of the mind—is some version of the anti-foundationalist assumption

---

54-67. The tenor of the ensuing debate in the mass media can be sampled in the Op/ed pages of the *New York Times*, May 18, 21, 24 and 26, 1996.

11. Nor will it do to conflate these "science wars" with the "culture wars" of William Bennet et al. First, unlike the transparent right-wing agenda of the culture wars, the most visible and sustained attacks from science have been carried out by scientists with avowed sympathies for the political left. One of the authors of *Higher Superstition* boasts impeccable leftist credentials, while Sokal explicitly argues that the jargon-filled speculation of *Social Text* represents the abandonment of serious political opposition by the academic left. Secondly, this skepticism is not confined to the mediocre practitioners of "normal science." The notion that, even if a naive realism is operative at the level of normal science, the undecidable, constructed, textual and irreducibly historical and political nature of reality will resonate with the most sophisticated minds in the field is belied by the inability of the Institute for Advanced Research to fill a post in Science Studies in three straight tries, rejecting successively Bruno Latour, Peter Galison and the physicist M. Norton Wise, the first and third by veto of the selection committee's recommendation. See "The Science Wars Flare at the Institute of Advanced Study," *The Chronicle of Higher Education* 43.36 (1997), A13.

that gestures to truth, meaning, rationality and an independent external world serve to mask the fact that scientific knowledge is the product of contingent, pragmatic, social processes that reflect the preoccupations and interest of its practitioners. The term "humanities-based critique" I have used is shorthand for that part of science studies, text-based and carried out largely in humanities departments, which brings to its engagement with science the style and preoccupations of contemporary literary theory. Science studies launched from a position in the social sciences, while sharing broad antifoundationalist assumptions with the humanities-based critique, differs in important ways. First, because of the ethnographic requirement for intensive fieldwork in the laboratory, the social sciences-based critique of science is carried out in contact with working scientists. Secondly, beyond the broad affinity provided by shared commitment to scientific methods, by some estimates upwards of half of researchers in the social sciences have bachelor's, master's, doctorate, or medical degrees in science, engineering, or mathematics.[12] Hence the science studies of Sharon Traweek, Evelyn Fox Keller, Andrew Pickering, Bruno Latour, and Steve Fuller, while often just as maddening to proponents of scientific realist orthodoxy, contain descriptions of scientific issues that are accepted by specialists.

Part of what makes humanities-based science studies easy to caricature is that a good proportion of its practitioners manifestly do not like or understand the object of their criticism. This seems like a frivolous observation, except that these are standard conditions for uncritical transference in scholarship.[13] Hence, this has suggested itself to a number of observers of the science wars as a criterion for distinguishing good and bad science studies, science studies which do, and do not, need to be taken seriously.

Of course, some of the research has been conducted by people who are hostile to science or ignorant about science, just as some anthropological fieldwork or historical archival work or literary interpretation is conducted by researchers who feel hostility to their subjects of inquiry, and some fieldwork and archival work and cultural interpretation is conducted by people who do not learn much about certain crucial practices of the people they

---

12. Traweek (1996), 132. Evelyn Fox Keller, for example, has a Ph.D. in physics.

13. Uncritical transference involves imputing to the object of study traits one refuses to recognize in oneself. See LaCapra (1985), 123-124.

study. This work is usually not too interesting to other researchers. (Traweek 1996, 132)[14]

Though the correlation with the division between the humanities and social sciences will not be exact, science seems likely to loom as a more clearly delineated "other" for a humanist with bad memories of math, than for a social scientist with a degree in the natural sciences. If the science wars indicates a broad opposition between foundation-alist and anti-foundationalist positions vis-a-vis scientific knowledge, the term "humanities-based science studies," points to a style based on the rhetoric of literary theory, and gestures to this potentially unproductive tension in relation to the object.

In the initial stage of the science wars, the attacks by Gross and Levitt, and Sokal were focused, and concentrated on the incompetent handling of scientific issues. This charge is justified, and falls harder on humanities-based writers. While being positioned myself in the humanities, I would like to follow Fox Keller, a social-scientist, and concede the initial point.

Stage Two: Retreat of the Scientists

It is in attempting to push the lines of debate beyond the initial, legitimate concern with standards of scholarly rigor in "certain precincts of the academy," however, to the anti-foundationalist assumptions of science studies in general, that the limitations of the science side have been revealed, faltering in the effort to move beyond the "second-rate" practitioners who were the easy target of *Higher Superstition*, and showing a tendency to fall back on metaphysical assumptions of its own. It is in this, though, that a structural problem in American intellectual debate is revealed that makes it a *requirement* to go outside this context to continue to pursue the issues legitimately raised.

Observers from outside the U. S. have largely been bemused by the ax-grinding and anathema of the public debate, with a common stance being to regard it as a "local skirmish" peculiar to the political and institutional situation of the American academy.[15] In *The Social*

14. See also Hacking (1999): "What is true is that many science-haters and know-nothings latch on to constructionism as vindicating their impotent hostility to the sciences. Constructionism provides a voice for that rage against reason" (67-68); and Holland (1994), cited above.

15. See Levidow, "Science Skirmishes and Science-Policy Research," (1996) for a view from Britain; Hacking (1999), vii-6, for views from Europe and Canada, and a

Construction of What?, however, Canadian Ian Hacking recognizes in the science wars a real problem about the conception of knowledge which all participants in the debate seem loathe to recognize. According to Hacking, it is precisely this failure to recognize their common ground that makes the participants talk past each other.

> Since neither scientists nor constructionists dare to use the word metaphysics, it is not surprising that they talk past each other, since each is standing on metaphysical ground in opposition to the other. Talk of metaphysics will seem, to many, a highbrow evasion of the issues current in the science wars. On the contrary, it is a central part of the story. (61)

Because it is indifferent to the question of the physical world, it makes immediate sense to ascribe a metaphysical character to the anti-foundationalist critique of science. But how can scientists be metaphysical? Here the distinction between science studies concerned with the natural sciences and the human sciences becomes important. The terrain on which the science wars is fought is, according to Hacking, only in part metaphysical. The purport of science studies in human sciences (the domain of categories like gender, childhood, madness, and youth homelessness) is moral, and is about unmasking or refuting the operation of power, oppression and ideology in the interaction between humans and the concepts and classifications science provides for them. However when science studies turns its sights to the natural sciences, to categories like quarks, tripeptides and load-raising machines, its purport is different (Hacking 1999, 57-61):

> Hence constructionism applied to the natural sciences was in the first instance metaphysical or epistemological—about pictures of reality or reasoning. (58)

Paradoxically, when pushed onto the real terms of the debate, to the question of foundations, it is scientists like Sokal, or Gross and Levitt who cannot be satisfied with the internal consistency of the conceptual system of physics, arrived at through experiment. They must posit a reality outside that system, literally a meta-physics to secure its truth. This takes the form of assertions by scientists about the ultimate character of reality not based on reason or evidence, evident as early as Sokal's revelation of the hoax in the pages of

---

special issue of the journal *Gendai Shisō* (Contemporary Thought) on "Science Wars" (1998) for a spectrum of opinions from Japan.

Lingua Franca. In attempting to press the point beyond the initial, limited purport of science and math literacy, Sokal reveals not the physicist's considered reflection on the way the history of his discipline as a collection of ideas relates to the physical world, but a naive objectivism:

> While my method was satirical, my motivation was utterly serious. What concerns me is the proliferation, not just of nonsense and sloppy thinking per se, but of a particular kind of nonsense and sloppy thinking: one that denies the existence of objective realities. . . . Intellectually, the problem with such doctrines is that they are false (when not simply meaningless). There *is* a real world; its properties are *not* merely social constructions/facts and evidence *do* matter. What sane person would contend otherwise? (Sokal 1996b, 63, italics in original)

Sokal's realism is naive for two reasons. First, it does not acknowledge legitimate skeptical arguments against the possibility of knowing a reality independent of our perception, present in philosophical traditions from Vedic and Buddhist to Greek, but put forth with the greatest clarity for the modern tradition in the eighteenth century by Berkeley. In *A Treatise Concerning the Principles of Human Knowledge*, Berkeley argued that we cannot know anything but what is present to our mind, that all that is present to our mind are impressions and concepts, and that the possibility that the cause of these impressions exists independent of our perception is strictly unknowable. This skeptical argument is unanswerable, arguably posed by the structure of sensation and perception in the vertebrate brain,[16] the twentieth century gloss being that what is unanswerable presupposes no question. But one of the ways of distinguishing modern philosophy from its more optimistic enlightenment predecessor is by the resolute attention given to the aporia into which rational thought seems inevitably to run. The domain on the other side may be conceived as the province of faith, mysticism or metaphysics with connotations

---

16. In a materialist account of the mind, sensation refers to the pattern of signals from sensory organs while perception involves the not at all obvious interpretation of these as signs from something in the external world. These would appear to have developed as two distinct kinds of mental representation in simple nervous systems. This is a common theme in contemporary cognitive neuroscience, and articles often contain a formulaic gesture in the opening paragraph to having solved the classical philosophical problems of materialism and idealism, or mind-body dualism. See Bownds (1999), 36-38.

pejorative or otherwise, but in thinkers from Kant to the early Wittgenstein it represents a domain to which the human intellect is un-
avoidably drawn in speculation, and about which science can have
nothing to say. In the call to order quoted above, Sokal is revealed to
revert to exactly the type of bald, unargued assertion he finds objec-
tionable in the object of his criticism.

Sokal's claim to know that there is a real world can also be criticized
on grounds of economy and falsifiability. One question to raise is
whether scientific realism adds something. I.e., given the coherence
and experimental success of scientific theories, does it add anything
by way of explanation to say, it is also like reality? One may also
invoke the criterion of scientific method and ask if the realist explanation is falsifiable. Is it possible to devise an experiment that could
demonstrate the opposite? This is the first sense in which the critics
of science studies fall back on their own metaphysical assumptions,
in the assertion of a reality independent of cognition, outside the
question of instrumentality and experiment, giving comforting guar-
antees of the truth of scientific knowledge.

The second way the putatively hard-headed realist position of
the science side gives evidence of a metaphysical character is in the
mystical affirmation of the rational character of the world. The dog-
matic or faith-based dismissal of certain radical possibilities in
twentieth-century physics is evident in the formal follow-up to *Higher
Superstition*, a detailed series of case studies by scientists, engineers
and traditional philosophers of science called *A House Built on Sand:
Exposing Postmodernist Myths about Science* (1998). This emerges in
particular when dealing with certain speculations by Bohr and Heisenberg still regarded as a central problem in physics. In a nice illustration
of the science studies doctrine that scientific facts harden with distance
from the communities that produced them, philosopher Paul Boghossian calls certain actively controversial statements of Bohr and Heisenberg on the philosophical implications of relativity and quantum
mechanics "naive" (Boghossian 1998, 24).

In a recap of the history of quantum physics for a twelve-part
celebration by *Science* magazine of the scientific achievements of the
twentieth century, Kleppner and Jackiw, both endowed professors of
physics at MIT write: "[N]ot only was quantum mechanics deeply
disturbing to its founders, today—75 years after the theory was essentially cast in its current form—some of the luminaries of science
remain dissatisfied with its foundations and its interpretation, even
as they acknowledge its stunning power" (893). In reference to the

somewhat different question of the extension to quantum field theory, "The procedure for quantizing the electromagnetic field that worked so brilliantly in QED has failed to work for gravity, in spite of a half-century of effort. The problem is critical." (See "One Hundred Years of Quantum Physics," *Science* (2000), 893-899).

It is important to specify what is meant by "radical possibilities" here. There is a facile notion, asserted with particular naivete in regard to quantum mechanics, relativity theory, and the emerging field of complexity theory, that counterintuitive developments in twentieth century math and science have invalidated classical descriptions in their domain, hence can be enlisted as an analogy for the politically committed problematization of a normative common sense in the humanities. The idea that Newtonian mechanics has been overturned and invalidated by the findings of twentieth-century physics is perhaps the central rhetorical miscalculation in the literary theoretical appropriation of science. Though it has radical implications for our understanding of the ultimate nature of matter, the corrections of quantum theory reduce effectively to zero in the domain of experience, leaving us free to live practically under classical assumptions. According to Werner Heisenberg, author of the famed Uncertainty Principle:

> So far as the concepts space, velocity, mass, etc. can be applied unhesitatingly—and that certainly applies to all experiences of everyday life—Newton's laws will be shown to be true. . . . How far this claim to validity can be justified can best be seen from the fact that Archimedes' laws of the simple lever still form today the basis of all load-raising machines and there can be no doubt they will do so for all time. (Heisenberg 1949, 42)

However, as Heisenberg patiently develops, there are highly unsettling implications about our conception of matter, and the relation of science to a unitary physical world at the outer reaches of the theory.[17] Boghossian is simply wrong. The more careful Levitt, however, acknowledges the problem.

> The two great achievements of twentieth century physics are relativity and quantum mechanics, each almost miraculously successful in its own predictive sphere. The problem is, they are

---

17. See Plotnitsky (1995), for an exhaustive exploration of the dynamics of this debate.

inconsistent with each other at a deep conceptual level; they both can't be 'right' in the literal sense. (Levitt 1998, 276)

Levitt admits here that the theories of physics as presently configured force one to entertain the possibility that the world at some fundamental level does not obey the Aristotelian laws of logic by which we construct our pictures of it, in particular the law of the excluded middle, and that the world we project from the scattered data of sense perception may prove ultimately to have no foundation. It may indeed be "a house built on sand." However, he downplays these radical possibilities with a simple will to optimism grounded neither in reason nor evidence: "My suggestion that they are solvable presupposes at least a modicum of hopefulness" (275). Levitt specifies the article of faith underlying this optimism, an enlightenment faith in progress, clearly stated and acknowledged:

> As I see it, the central, long-term project of theoretical physics is to produce a coherent, foundational point of view, one that not only unites (perhaps with subtle modifications) relativity and quantum mechanics but also provides a model that dispels the seeming paradoxes that have bedeviled our understanding of the latter. (276)

It is a principle of mathematics that one strives for economy and coherence in a conceptual system, however it is an article of faith, that cannot be proved by any number of past reductivist successes, that the world will display the same economy. Like Einstein in his disputes with Bohr, what Levitt would like is to secure a restricted economy from the general economy of twentieth-century physics. This is a matter of faith, hinted at in the idea of paradoxes that bedevil, and labelled clearly as such in Einstein's summary "God doesn't play dice with the universe." In the mystical affirmation of the rational character of the world, the critique of science studies initiated by Gross and Levitt and their followers runs aground on its own metaphysical suppositions. That is to say, in the effort to refute science studies, rather than just goad the opposition into more effective practice, the argument moves onto territory that is other to science.

With the failure of *A House Built on Sand* to deliver the refutation of first-rate science studies hinted at in the turkeyshoot of *Higher Superstition*, and the retreat of principal figures to unassailable metaphysical ground, the science wars was bound to end, as with its predecessor the Two Cultures debate, in an impasse. Styling himself a "foreign correspondent," Hacking gives perhaps the best description of what it felt like to witness the descent into spectacle: "[Participants] won't talk to each other, or else they talk past each other, because one side is so contentiously 'constructionist' while the other is so dismissive of the idea. In larger arenas, public scientists shout at sociologists, who shout back. You almost forget there are issues to discuss" (Hacking 1999, vii).

Stage Three: Back to Where We Started

The deeper problem revealed here is the repetition of the Two Cultures debate. That the same set of issues would come up in substantially the same form 35 years later—with participants clashing emotionally and withdrawing without any resolution or even indication of which problems had been elucidated and given organization, and which remained to be considered—indicates that there is a genuine issue, and that something is preventing the issue from being engaged. Though the peculiarly public and acrimonious nature of the science wars seem to mark it as an anomaly in an academic context, there is an increasing consciousness among commentators that this type of dynamic, where "exchanges between proponents of rival views lead recurrently to deadlock and impasse" (Herrnstein-Smith 1997, ix), may rather represent something typical in the political and intellectual configurations of academic debate in North America. This quality of impasse itself then, rather than the content of particular disputes, emerges as an object of inquiry.

Barbara Herrnstein-Smith's *Belief and Resistance: Dynamics of Contemporary Intellectual Controversy* (1997) discusses in detail a variety of contemporary controversies over questions of evidence in law, of the relation between organism and environment in biology, and of the social construction of scientific facts in science studies, and finds the same core opposition between anti-foundationalist arguments and foundationalist assumptions structuring virtually every field of intellectual inquiry in North America today. Herrnstein-Smith locates the key to the particular dynamics of contemporary American discourse, though, in a tendency to "epistemic self-privileging" on the part of both parties to dispute, a mode of discourse she calls asymmetrical. This adversarial dynamic consists

> [n]ot simply in preferring your own judgements or beliefs to those of other people (a self-privileging that is not, in itself, theoretically problematic), but in maintaining that you and the members of your group prefer your judgements or beliefs because they are objectively correct. . . . while other people prefer their

differing judgements or beliefs because those people are benighted or otherwise deficient. (7-8)

If the barrenness and sterility of the science wars turns out to be not an accidental feature or failure of particular participants that can be remedied by more rigorous application of the protocols of academic discourse, but a structural feature of intellectual controversy in the North American academy, then it makes sense to step outside this dynamic to get some purchase on the legitimate issues raised by the science wars. To frame it in Herrnstein-Smith's words, what we find in contemporary U. S. academic discourse on the relation between literature and science is an asymmetrically framed dispute. What is required to pursue the question of literature and science is a symmetrically framed dispute. Herrnstein-Smith imagines such a discourse as follows:

> But the differences could also be framed *symmetrically* on both sides as reflections of Our/Their differences of conceptual style and cognitive taste, differences that would themselves be seen as products of Our/Their more or less extensive differences of individual temperament and intellectual history, as played out within more or less different disciplinary cultures and sustained under more or less different epistemic conditions. (Herrnstein-Smith 1997, 136-137)

That is what I have searched for in the unique intellectual milieu of early twentieth-century Japan. It is the contention of this study that the centrality of the literature-science problematic for foundational figures like Mori Ogai (1862-1922) and Natsume Sōseki (1867-1916), the purposefulness with which they pursued interdisciplinary careers and their commitment to take seriously the claims of both sides reflect not merely personal proclivities, but the institutional conditions for a symmetrically framed dispute, and that the legacy of this still sounds in Japanese intellectual life today. But to partake, one first needs to recognize the difference.

Bibliography

Arima, Akito. 1995. Umarete kuru no wa gojūnen hayakatta. *Asahi Kagaku* 55 (11):24-28.

———. 1996. Kagaku to bungaku ga deau toki. *Front* 9 (3):17-20.

Asada Akira, and Fukuda Kazuya. 1999. 'Nihon' wo koeru kyōiku. *Bungei Shunjū* (6):139-147.

Baldick, Chris. 1990. *The Concise Oxford Dictionary of Literary Terms*. Oxford: Oxford University Press.

Bartholomew, James R. 1989. *The Formation of Science in Japan*. New Haven: Yale University Press.

Bataille, Georges. 1988. *The Accursed Share: An Essay on General Economy*. Translated by R. Hurley. v. 1. New York: Zone Books.

Benjamin, Walter. 1968. *Illuminations*. New York: Shocken Books.

Berkeley, George. 1952. A Treatise Concerning the Principles of Human Knowledge. In *Locke, Berkeley, Hume*. Chicago: Encyclopedia Britannica.

Berlinski, David. 1995. *A Tour of the Calculus*. New York: Vintage Books.

Bhaskar, Roy. 1979. *The Possibility of Naturalism: A Philosophical Critique of the Contemporary Human Sciences, Critical Realism: Interventions*. London: Routledge.

Block, Edwin F. 1993. *Rituals of Dis-Integration: Romance and Madness in the Victorian Psychomythic Tale*. New York: Garland.

Blumenberg, Hans. 1987. *The Genesis of the Copernican World*. Cambridge: MIT Press.

Boghossian, Paul A. 1998. What the Sokal Hoax Ought to Teach Us. In *A House Built on Sand: Exposing Postmodernist Myths about Science*, edited by N. Koertge. New York: Oxford University Press.

Bordwell, David. 1989. *Making Meaning: Inference and Rhetoric in the Interpretation of Cinema*. Cambridge: Harvard University Press.

Bownds, M. Deric. 1999. *The Biology of Mind: Origins and Strucutres of Mind, Brain, and Consciousness*. Bethesda, MD: Fitzgerald Science Press.

Calvin, William. 1999. Review of Antonio Damasio's *The Feeling of What Happens: Body and Emotion in the Making of Consciousness*. New York Times Book Review, 24 October 1999, 8.

Campbell, Paulette W. 1999. NEH Official Calls for Broader Research into U. S. Humanities Policy. *The Chronicle of Higher Education* (June 25):A42-A43.

Chalmers, David J. 1996. *The Conscious Mind: In Search of a Fundamental Theory*. New York: Oxford University Press.

Chinn, W. G., and N. E. Steenrod. 1966. *First Concepts of Topology*. New York: Random House.

Christie, Dan E. 1976. *Basic Topology*. New York: Macmillan Publishing Co., Inc.

Conant, James. 1991. On Bruns, On Cavell. *Critical Inquiry* 17 (3):616-634.

Coughlan, G. D., and J. E. Dodd. 1991. *The Ideas of Particle Physics: An Introduction for Scientists*. Cambridge: Cambridge University Press.

Crary, Jonathan. 1988. Modernizing Vision. In *Vision and Visuality*, edited by H. Foster. Seattle: Bay Press.

Davis, Peter. 1994. Working Chaos. *Gendai Shisō* 22 (6):116-129.

Deleuze, Gilles. 1993. *The Fold: Leibniz and the Baroque*. Translated by T. Conley. Minneapolis: University of Minnesota Press.

Derrida, Jacques. 2001. "The Future of the Profession, or the Unconditional University." Public lecture at University of Florida, Gainesville, 12 April.

Eagleton, Terry. 1983. *Literary Theory: An Introduction*. Minneapolis: University of Minnesota Press.

Edogawa, Rampo. 1956. *Japanese Tales of Mystery and Imagination*. Trans. by J. B. Harris. Tokyo: Tuttle Co.

———. 1991. *Edogawa Rampo*. v. 19, *Nihon Bungaku Zenshū*. Tokyo: Chikuma Shobō.

Eto, Jun. 1970. *Sōseki to sono jidai*. v. 2. Tokyo: Shinchōsha.

Eysenck, H. J. 1995. *Genius: The Natural History of Creativity*. Cambridge: Cambridge University Press.

Ezawa, Hiroshi. 1997. Kaisetsu—Torahiko no butsurigaku kan. In *Terada Torahiko Zenshū*. Tokyo: Iwanami Shoten.

Feynman, Richard. 1979. *Interview*: <http://www.omnimag.com/archives/interviews/feynman.html>.

———. 1995. *Six Easy Pieces: Essentials of Physics Explained by Its Most Brilliant Teacher*. Reading, MA: Helix Books.

Forster, Michael N. 1989. *Hegel and Skepticism*. Cambridge: Harvard University Press.

———. 1998. *Hegel's Idea of a Phenomenology of Spirit*. Chicago: University of Chicago Press.

Fox Keller, Evelyn. 1996. The Sokal Hoax: A Forum. *Lingua Franca* July/August:58-64.

Fox-Keller, Evelyn. 1998. Kuhn, Feminism and Science? *Configurations* 6 (1):15-21.

Franklin, Sarah. 1996. Making Transparencies: Seeing Through the Science Wars. *Social Text* 46/47 (Spring/Summer):141-155.

Freeman, Walter J., and Christine A. Skarda. 1987. How Brains Make Chaos in Order to Make Sense of the World. *Behavioral and Brain Sciences* 10 (2):161-195.

Fuller, Steve. 1996. Does Science Put an End to History, or History to Science? *Social Text* 46/47 (Spring/Summer):27-42.

———. 1997. *Science, Concepts in Social Thought*. Minneapolis: University of Minnesota Press.

Golinski, Peter. 1998. *Making Natural Knowledge: Constructivism and the History of Science*. Cambridge: Cambridge University Press.

Gombrich, Ernst. 1960. *Art and Illusion*. Princeton: Princeton University Press.

Gotō, Akio, and Akimasa Kanno. 1995. Shōsetsu no toporojii. *Gunzō* 11:142-169.

Graff, Gerald. 1995. Determinacy/Indeterminacy. In *Critical Terms for Literary Study*, edited by F. Lentricchia and T. McLaughlin. Chicago: University of Chicago Press.

Greenberger, Daniel. 2000. Bohr the Innovator? Or Bohr the Intimidator? *Science* 287 (24 March):2166-2167.

Gross, Paul R., and Norman Levitt. 1994. *Higher Superstition: The Academic Left and Its Quarrels with Science*. Baltimore: The Johns Hopkins University Press.

Guillory, John. 1993. *Cultural Capital: The Problem of Literary Canon Formation*. Chicago: University of Chicago Press.

Hacking, Ian. 1995. *Rewriting the Soul: Multiple Personality and the Sciences of Memory*. Princeton: Princeton University Press.

———. 1999. *The Social Construction of What?* Cambridge: Harvard University Press.

Haraway, Donna. 1989. *Primate Visions: Gender, Race and Nature in the World of Modern Science.* New York: Routledge.

Harvey, David. 1990. *The Condition of Postmodernity.* Oxford: Blackwell Publishers.

Hayles, N. Katherine. 1990. Self-Reflexive Metaphors in Maxwell's Demon and Shannon's Choice. In *Literature and Science: Theory and Practice,* edited by S. Peterfreund. Boston: Northeastern University Press.

Heath, Stephen. 1983. Le Pere Noël. *October* 26 (Fall):63-115.

Heidegger, Martin. 1971. *Poetry Language, Thought.* Translated by A. Hofstadter. New York: Harper & Row.

Heilman, Richard. 2000. Creativity and CNS. Gainesville: University of Florida.

Heisenberg, Werner. 1949. *Philosophical Problems of Nuclear Science.* London: Faber & Faber.

Herrnstein-Smith, Barbara. 1997. *Belief and Resistance: The Dynamics of Contemporary Intellectual Controversy.* Berkeley: University of California Press.

Holden, Arun V., ed. 1986. *Chaos.* Princeton: Princeton University Press.

Holland, Norman. 1994. Reader Response Already Is Cognitive Criticism. *Stanford Humanities Review* 4 (1):65-66.

———. 2000. Creativity and the Stock Market. *Bulletin of Psychology and the Arts* 1 (2):62-64.

Ikeda Kensuke. 1994. Fukuzatsukei no écriture. *Gendai Shisō* 22 (6):160-171.

Ingarden, Roman. 1973. *The Literary Work of Art: An Investigationon the Borderlines of Ontology, Logic, and Theory of Literature.* Translated by G. C. Grabowicz. Evanston: Northwestern University Press.

Isozaki Arata. 1999. Kenchiku in okeru 'Nihontekina mono'. *Hihyō Kūkan* II (21):186-204.

James, William. 1952. *The Principles of Psychology.* v. 53, *Great Books of the Western World.* Chicago: Encyclopedia Britannica.

Jameson, Fredric. 1993. Postmodernism, or the Cultural Logic of Late Capitalism. In *Postmodernism: A Reader,* edited by T. Docherty. New York: Columbia University Press.

Jasny, Barbara, and Phil Szuromi. 2001. This Week in Science. *Science* 291 (5507):1155-1157.

Johnson, Steven. 1996. Strange Attraction. *Lingua Franca* Mar/Apr:42-50.

Joy, Bill. 2000. Why the Future Doesn't Need Us. *Wired* April:238-262.

Kamei, Hideo. 2003. *Transformations of Sensibility: The Phenomenology of Meiji Literature.* Translated by Bourdaghs et al. Ann Arbor: Michigan Monograph Series in Japanese Studies.

Kaneko Kunihiko. 1995. Kaos to wa nanika. In *Chi no ronri,* edited by Funabiki Takeo. Tokyo: University of Tokyo Press.

———. 1996. Fukuzatsukei: Kaosuteki shinario kara seimeiteki shinario e. *Gendai Shisō* 24 (13):79-86.

Karatani, Kōjin. 1991. Uchigawa kara mita sei. In *Natsume Sōseki.* Tokyo: Shōgakkan.

———. 1993. *The Origins of Modern Japanese Literature.* Translated by B. deBary, et al. Durham: Duke University Press.

———. 1995. *Architecture as Metaphor: Language, Number, Money.* Cambridge: MIT Press.

———. 1997. Bigaku no kōyō—'Orientarizumu' ikō. *Hihyō Kūkan* II (14):42-55.

———. 2001. Buddhism, Marxism and Fascism in 1920's and 1930's Discourse in Japan. In *New Critical Perspectives in Twentieth Century Japanese Thought,* edited by L. Monnet. Montreal: University of Montreal Press:185-223.

Karatani, Kojin, Akira Asada, and Hiroki Azuma. 1998. Transcritique to (shite no) datsukōchiku II-18 (1998). *Hihyō Kūkan* II (18):2-34.

Karatani, Kōjin, Yōichi Komori, and Teruhiko Tsuge. 1991. 'Hihyō' to wa nanika. *Kokubungaku* Heisei 3 (6):6-27.

Kato, Shuichi. 1979. *A History of Modern Japanese Literature: The Modern Years.* v. 3. Tokyo: Kodansha.

Kawakami, Shinichi. 1996. Zen-chikyū shi kaidoku keikaku— Kōkōgaku kara no shiza. *Gendai Shisō* 24 (3):210-220.

Kenny, Anthony. 1994. Afterword. In *Oxford History of Western Philosophy.* New York: Oxford University Press.

Kitagawa, Joseph M. 1987. *On Understanding Japanese Religion.* Princeton: University of Princeton Press.

Kleppner, Daniel, and Roman Jackiw. 2000. One Hundred Years of Quantum Physics. *Science* 289 (11 Aug):893-899.

Kobayashi Hideo. 1995. *Literature of the Lost Home: Literary Criticism 1924-1939.* Translated by P. Anderer. Stanford: Stanford University Press.

Kōchi Kenritsu Bungakkan. 1986. *Terada Torahiko jikkenshitsu e.* Kōchi.

Komiya, Toyoyuki. 1966. *Bungakuron* kaisetsu. In *Sōseki Zenshū.* Tokyo: Iwanami Shoten.

Komori, Yōichi. 1995. *Sōseki o yominaosu*. v. 37, *Chikuma Shinsho*. Tokyo: Chikuma Shobō.

Kōsaka, Masaaki. 1958. *Japanese Thought in the Meiji Era*. Translated by D. Abosch. v. IX, *Japanese Culture in the Meiji Era*. Tokyo: Pan-Pacific Press.

Koyama, Keita. 1991. *Sōseki ga mita butsurigaku*. v. 1053, *Chūkō Shinsho*. Tokyo: Chuōkōronsha.

———. 1996. 'Terada butsurigaku' no sugao. *Front* 9 (3 (Terada Torahiko tokushū)):12-14.

Kumazawa, Mineo, and Keisuke Itō. 1996. Kōkōgaku to wa nanika. *Gendai Shisō* 24 (3):124-149.

Kuhn, Thomas S. 1970. *The Structure of Scientific Revolutions*. Chicago: University of Chicago Press.

LaCapra, Dominick. 1985. *History and Criticism*. Ithaca: Cornell University Press.

Leach, Edmund. 1976. *Culture and Communication: The Logic by Which Symbols Are Connected*. Cambridge: Cambridge University Press.

Leavis, F. R. 1972. Two Cultures? The Significance of Lord Snow. In *Nor Shall My Sword: Discourses on Pluralism, Compassion and Social Hope*. New York: Barnes & Noble.

Lehman, Peter. 1987. The Mysterious Orient, the Crystal Clear Orient, the Non-Existent Orient: Dilemmas of Western Scholars of Japanese Film. *Journal of Film and Video* xxxix (Winter):5-15.

Levidow, Les. 1996. Science Skirmishes and Science-Policy Research. *Social Text* 46/47:199-206.

Levine, George. 1996. What is Science Studies for and Who Cares? *Social Text* 46/47:113-128.

Levitt, Norman. 1998. The End of Science, the Central Dogma of Science Studies. In *A House Built on Sand: Exposing Postmodernist Myths about Science*, ed. N. Koertge. New York: Oxford University Press.

Leyton, Michael. 2001. *A Generative Theory of Shape*. Berlin: Springer-Verlag.

Ludwig, Arnold M. 2000. Fractals, Madness and Creative Achievement. *Bulletin of Psychology and the Arts* 1 (2):47-48.

Lyotard, Jean-Francois. 1983. *The Postmodern Condition: A Report on Knowledge*. Minneapolis: University of Minnesota Press.

Maeda Ai. 1982. *Toshi Kūkan no naka no bungaku*. Tokyo: Chikuma Shobō.

——. 1989. *Kindai dokusha no seiritsu*. v. 2, *Maeda Ai Chosakushū*. Tokyo: Chikuma Shobō.

Mandelbrot, Benoit. 1983. *The Fractal Geometry of Nature*. New York: W. H. Freeman and Co.

Martindale, Colin. 1981. *Cognition and Consciousness*. Homewood, IL: Dorsey.

——. 2000. Disinhibition, Dopamine and Creativity. *Bulletin of Psychology and the Arts* 1 (2):49-53.

Matsui, Sakuko. 1975. *Natsume Sōseki as a Critic of English Literature*. Tokyo: Centre for East Asian Cultural Studies.

Matsumura, Tatsuo, ed. 1971. *Natsume Sōseki shū I*. v. 24, *Nihon Kindai Bungaku Taikei*. Tokyo: Kadokawa Shoten.

Matsuno, Kōichirō, and Ichirō Tsuda. 1996. Fukuzatsukei no shinario. *Gendai Shisō* 24 (13):50-78.

Matteuzzi, Maurizio. 1995. Why AI Is Not A Science. *Stanford Humanities Review* 4 (2):205-219.

Mervis, Jeffrey. 2000. 2001 Budget: Spending Bills Show No Sign of Surplus—Yet. *Science* 289 (7 July):31.

Miyoshi, Masao. 1974. *Accomplices of Silence: The Modern Japanese Novel*. Berkeley: University of California Press.

Murakami, Yasusuke. 1988. The Debt Comes Due for Mass Higher Education. *Japan Echo* XV (3):71-80.

Muramatsu Sadataka, et.al., ed. 1983. *Izumi Kyōka*. Tokyo: Ofūsha.

Muraoka, Isamu. 1975. *Sōseki shiryō—Bungakuron nōto*. Tokyo: Iwanami Shoten.

Nakayama, Shigeru. 1997. Torahiko paradigm. *Terada Torahiko Zenshū Geppō* 2:1-3.

Natsume, Sōseki. 1992. *Kokoro and Selected Essays*. Translated by J. Rubin. Lanham: Madison Books.

Newton, Isaac. 1952. Mathematical Principles of Natural Philosophy. In *Newton, Huygens*. Chicago: Encyclopedia Britannica.

Obayashi, Shinji, and Toshiteru Morita. 1994. *Kagaku shisō no keifugaku*. Tokyo: Minerva Shobō.

Ogai, Mori. 1994. *Youth & Other Stories*. Translated by R. Bowring. Edited by J. T. Rimer. Honolulu: University of Hawaii Press.

Ogawa, Tōru. 2000. Introduction to the Symposium on the Science of Form: The Concept of the Symposium, edited by K. Fushimi, K. Miura, T. Masunari and D. Nagy. <http://kafka.bk.tsukuba.ac.jp/kus/Introduction/Introduction.htm>.

Ogawa, Tōru, Koji Fushimi, Koryo Miura, Takashi Masunari, and Denes Nagy, eds. 1996. *Katachi U Symmetry*: Springer Verlag.

Oguma, Eiji. 1998. Terada Torahiko: Rinkai ni kirikomu shunpatsuryoku. *Asahi Shinbun*, September 24: 17.

Okuma, Shigenobu, ed. 1909. *Fifty Years of New Japan*. 2 vols. New York: E. P. Dutton.

Ota Bumpei. 1971. *Terada Torahiko: Sono sekai to ningenzō*. Tokyo: Toshi Shuppansha.

Palmer, Richard. 1969. *Hermeneutics*. Evanston: Northwestern University Press.

Paxson, James. 1997. The Institutional Trajectories of Modern Science and Contemporary Critical Theory. Paper read at International Conference on Narrative, at Gainesville, FL.

Pickering, Andrew. 1995. Concepts and the Mangle of Practice: Constructing Quaternions," in SAQ, special issue on "Mathematics, Science and Postclassical Theory," v. 94, #2 (Spring).

Pinker, Steven. 1994. *The Language Instinct: How the Mind Creates Language*. New York: Harper Collins.

Plotnitsky, Arkady. 1994. *Complementarity: Anti-Epistemology After Bohr and Derrida*. Durham: Duke University Press.

Poulet, Georges. 1966. *The Metamorphoses of the Circle*. Translated by C. Dawson. Baltimore: The Johns Hopkins University Press.

Pullen, Keats A. 1962. *Theory and Application of Topological and Matrix Methods*. New York: John F. Rider Publisher, Inc.

Pyenson, Lewis. 1981. History of Physics. In *Encyclopedia of Physics*, edited by R. G. Lerner and G. L. Trigg. New York: Addison Wesley.

Quine, W. V. 1995. *From Stimulus to Science*. Cambridge: Harvard University Press.

Ray, William. 1984. *Literary Meaning: From Phenomenology to Deconstruction*. Cambridge: Basil Blackwell.

Rickman, Hans Peter. 1996. *Philosophy in Literature*. Madison: Associated University Presses.

Roden, Donald. 1980. *Schooldays in Imperial Japan*. Berkeley: University of California Press.

Rotman, Brian. 1994. Trapped in Hypostases. *Stanford Humanities Review* 4 (1):99-101.

Ruelle, Brian. 1992. *Chance and Chaos*. Princeton: Princeton University Press.

Saionji, Kimmochi. 1909. National Education in the Meiji Era. In *Fifty Years of New Japan*, edited by Okuma Shigenobu. New York: E. P. Dutton.

Saito Bunichi. 1991. *Miyazawa Kenji—Shijigenron no tenkai*. Tokyo: Kokubunsha.

Saitō, Nobusaku. 1907. Kyōka to romanchikku. *Taiyō* S 40 (9):unpaginated.

Sakai Naoki. 1990. 'Riron' to sono 'Nihonteki' seiyaku. *Bungaku* Fall:64-73.

———. 2000. Subject and Substratum: On Japanese Imperial Nationalism. *Cultural Studies* 14 (3/4):462-530.

Sasaki, Chikara. 1996. *Kagakuron nyūmon*. v. 457, *Iwanami Shinsho*. Tokyo: Iwanami Shoten.

Scruton, Roger. 1994. Continental Philosophy from Fichte to Sartre. In *Oxford History of Western Philosophy*, edited by A. Kenny. Oxford: Oxford University Press.

Serres, Michel, and Bruno Latour. 1995. *Conversations on Science, Culture, and Time*. Translated by R. Lapidus. Ann Arbor: University of Michigan Press.

Shaffer, Elinor S., ed. 1998. *The Third Culture: Literature and Science*. v. 9, *European Cultures: Studies in Literature and the Arts*. Berlin: Walter de Gruyter.

Shimada, Atsushi. 1960. Sōseki no shisō. *Bungaku* October.

———. 1961. Sōseki no shisō 2. *Bungaku* February.

Silverman, Kaja. 1986. Suture (Excerpts). In *Narrative, Apparatus, Ideology*, edited by P. Rosen. New York: Columbia University Press.

Sikivie, Pierre. 1996. The Pool-Table Analogy with Axion Physics. *Physics Today* December:22-27.

Skarda, Christine A., and Walter J. Freeman. 1987. How Brains Make Chaos in Order to Make Sense of the World. *Behavioral and Brain Sciences* 10 (2):161-195.

Snow, C. P. 1959. *The Two Cultures and the Scientific Revolution*. New York: Cambridge University Press.

Sokal, Alan D. 1996. A Physicist Experiments with Cultural Studies. *Lingua Franca* May/June:62-64.

———. 1996. Transgressing the Boundaries: Toward a Transformative Hermeneutics of Quantum Gravity. *Social Text* 45/46:217-252.

Sōseki, Natsume. 1961. *I Am a Cat*. Translated by S. Katsue. Tokyo: Kenkyūsha.

———. 1966. *Natsume Sōseki Zenshū*. 17 vols. Tokyo: Iwanami Shoten.

———. 1978. *And Then.* Translated by N. Field, *Michigan Classics in Japanese Studies.* Ann Arbor: University of Michigan Press.

Stwertka, Albert. 1996. *Guide to the Elements.* New York: Oxford University Press.

Takahashi, Michiko. 1966. Natsume Sōseki no *Bungakuron* ni okeru Ribō no *Kanjō no Shinrigaku.* In *Bungei Kenkyū.* Tokyo: Nihon Bungei Kenkyūkai.

Takaki, Ryuji. 1997. Terada Torahiko to katachi no kagaku. *Terada Torahiko Zenshū* Geppō 10:1-4.

Takaki, Ryuji, Yoji Arai, and Masaki Utsumi. 2000. Promotion of the Morphological Sciences. <http://kafka.bk.tsukuba.ac.jp/kus/takaki.htm>.

Teller, Edward. 1996. *Why Things Bite Back: Technology and the Revenge of Unintended Consequences.* New York: Vintage Books.

Terada, Torahiko. 1939. *Scientific Papers.* 6 vols. Tokyo: Iwanami Shoten.

———. 1991. *Terada Torahiko Zenzuihitsushū.* 6 vols. Tokyo: Iwanami Shoten.

———. 1996. *Terada Torahiko Zenshū.* 30 vols. Tokyo: Iwanami Shoten.

Thompson, D'Arcy Wentworth. 1992. *On Growth and Form.* New York: Dover Publications.

Todorov, Tzvetan. 1985. 'Race,' Writing, and Culture. In *'Race,' Writing, and Difference,* edited by H. L. Gates. Chicago: University of Chicago Press.

Tokushū: Saiensu Waazu. 1998. *Gendai Shisō* 26 (13).

Traweek, Sharon. 1988. *Beamtimes and Lifetimes: the World of High Energy Physicists.* Cambridge: Harvard University Press.

———. 1996. Unity, Dyads, Triads, Quads, and Complexity: Cultural Choreographies of Science. *Social Text* 45/46: 129-139.

Valdés, Mario, and Étienne Guyon. 1998. Serendipity in Poetry and Physics. In *The Third Culture: Literature and Science,* edited by E. S. Shaffer. Berlin: Walter de Gruyter.

Van Rees, C. J. 1983. How a Literary Work Becomes a Masterpiece: On the Threefold Selection Practised by Literary Criticism. *Poetics* 12:397-417.

Venkataramani, Shankar C. 1997. *The Chaos Revolution and Beyond: Physics in a Nonlinear World.* September 27-December 6, *Arthur H. Compton Lectures.* University of Chicago.

Vickers, Steven. 1989. *Topology Via Logic.* Cambridge: Cambridge University Press.

Virilio, Paul. 1994. *The Vision Machine.* London: BFI Publishing.

Waldrop, Mitchell M. 1992. *Complexity: The Emerging Science at the Edge of Chaos and Order*. New York: Simon & Schuster.

Weingart, Peter, and Sabine Maasen. 1997. The Order of Meaning: The Career of Chaos as a Metaphor. *Configurations* 5:463-520.

Weiss, Barbara. 1986. *The Hell of the English: Bankruptcy and the Victorian Novel*. Lewisburg, PA: Bucknell University Press.

White, Eric Charles. 1990. Contemporary Cosmology and Narrative Theory. In *Literature and Science: Theory and Practice*, edited by S. Peterfreund. Boston: Northeastern University Press.

Wiener, Norbert. 1950. Progress and Entropy. In *The Human Use of Human Beings*. London: Eyre & Spottiswoode.

Williams, Raymond. 1976. *Keywords: A Vocabulary of Culture and Society*. New York: Oxford University Press.

Winograd, Terry, and Fernando Flores. 1985. *Understanding Computers and Cognition: A New Foundation for Design*. Norwood, NJ: Ablex.

Yoshimi, Shunya. 1991. Toporojii no henyō. In *Hōhō toshite no kyōkai*, edited by N. Akasaka. Tokyo: Shinyōsha.

Yoshinaga, Norimasu. 1996. Fukuzatsukei to wa nanika? *Gendai Shisō* 24 (13):87-102.

Yoshioka, Hiroshi. 1996. Risei no "fuchi" e. *Gendai Shisō* 24 (13):190-203.

Young, Robert. 1990. *White Mythologies: Writing History and the West*. London: Routledge.

Zaldarriaga, Matias, and David W. Hogg. 2000. The Big Bang's Radical Brother. *Science* 290: 2079-80.

Zizek, Slavoj. 1989. *The Sublime Object of Ideology*. London: Verso.

Index

CORNELL EAST ASIA SERIES

88 Chang Soo Ko, tr., *Sending the Ship Out to the Stars: Poems of Park Je-chun*

89 Thomas Lyons, *The Economic Geography of Fujian: A Sourcebook*, Vol. 2

90 Brother Anthony of Taizé, tr., *Midang: Early Lyrics of So Chong-Ju*

91 Chifumi Shimazaki, *Battle Noh: Parallel Translations with Running Commentary*

92 Janice Matsumura, *More Than a Momentary Nightmare: The Yokohama Incident and Wartime Japan*

93 Kim Jong-Gil, tr., *The Snow Falling on Chagall's Village: Selected Poems of Kim Ch'un-Su*

94 Wolhee Choe & Peter Fusco, trs., *Day-Shine: Poetry by Hyon-jong Chong*

95 Chifumi Shimazaki, *Troubled Souls from Japanese Noh Plays of the Fourth Group*

96 Hagiwara Sakutarō, *Principles of Poetry* (Shi no Genri), tr. Chester Wang

97 Mae J. Smethurst, *Dramatic Representations of Filial Piety: Five Noh in Translation*

98 Ross King, ed., *Description and Explanation in Korean Linguistics*

99 William Wilson, *Hōgen Monogatari: Tale of the Disorder in Hōgen*

100 Yasushi Yamanouchi, J. Victor Koschmann and Ryūichi Narita, eds., *Total War and 'Modernization'*

101 Yi Ch'ŏng-jun, *The Prophet and Other Stories*, tr. Julie Pickering

102 S.A. Thornton, *Charisma and Community Formation in Medieval Japan: The Case of the Yugyō-ha (1300-1700)*

103 Sherman Cochran, ed., *Inventing Nanjing Road: Commercial Culture in Shanghai, 1900-1945*

104 Harold M. Tanner, *Strike Hard! Anti-Crime Campaigns and Chinese Criminal Justice, 1979-1985*

105 Brother Anthony of Taizé & Young-Moo Kim, trs., *Farmers' Dance: Poems by Shin Kyong-nim*

106 Susan Orpett Long, ed., *Lives in Motion: Composing Circles of Self and Community in Japan*

107 Peter J. Katzenstein, Natasha Hamilton-Hart, Kozo Kato, & Ming Yue, *Asian Regionalism*

108 Kenneth Alan Grossberg, *Japan's Renaissance: the Politics of the Muromachi Bakufu*

109 John W. Hall & Toyoda Takeshi, eds., *Japan in the Muromachi Age*

110 Kim Su-Young, Shin Kyong-Nim, Lee Si-Young; *Variations: Three Korean Poets*; Brother Anthony of Taizé & Young-Moo Kim, trs.

111 Samuel Leiter, *Frozen Moments: Writings on Kabuki, 1966-2001*

112 Pilwun Shih Wang & Sarah Wang, *Early One Spring: A Learning Guide to Accompany the Film Video* February

113 Thomas Conlan, *In Little Need of Divine Intervention: Scrolls of the Mongol Invasions of Japan*

114 Jane Kate Leonard & Robert Antony, eds., *Dragons, Tigers, and Dogs: Qing Crisis Management and the Boundaries of State Power in Late Imperial China*

FORTHCOMING

Order online: www.einaudi.cornell.edu/eastasia/CEASbooks, or contact Cornell East Asia Series Distribution Center, 95 Brown Road, Box 1004, Ithaca, NY 14850, USA; toll-free: 1-877-865-2432, fax 607-255-7534, ceas@cornell.edu

4-04/.7M pb/SB